Art & Science

A CURRICULUM FOR K–12 TEACHERS FROM THE J. PAUL GETTY MUSEUM

Art & Science
A Curriculum for K–12 Teachers
from the J. Paul Getty Museum

The J. Paul Getty Museum
Los Angeles

Published by the J. Paul Getty Museum
Getty Publications
1200 Getty Center Drive, Suite 500
Los Angeles, California 90049–1682
www. getty.edu/publications

Elizabeth S. G. Nicholson, Editor
Catherine Lorenz, Designer
Suzanne Watson and Pamela Heath, Production Coordinators
Printed in China

Library of Congress Cataloging-in-Publication Data

Art & science : a curriculum for K–12 teachers from the J. Paul Getty Museum.
p. cm.
ISBN 978-1-60606-141-1 (pbk.)
1. Art and science—Study and teaching (Elementary)—California. 2. Art and science—Study and teaching (Elementary)—United States. 3. Art and science—Study and teaching (Secondary)—California. 4. Art and science—Study and teaching (Secondary)—United States. 5. Teaching—Aids and devices. I. J. Paul Getty Museum. Education Dept. II. Title: Art and science.
N72.S3A645 2013
701'.05—dc23

2012041648

Note to readers: Dimensions of art objects are given as height by width or as height by width by depth.

Cover: Jan van Kessel, *Butterfly, Caterpillar, Moth, Insects, and Currants* *(detail; see page 7).*

Contents

Acknowledgments

The J. Paul Getty Museum Education staff wrote this curriculum with input from Getty Museum conservators and curators. In addition, the curriculum was created under the guidance of a teacher advisory group. For their assistance in this project, we would like to thank Tyson Evans, John Muir Middle School; Kimberly Garcia, San Jose Street Elementary School; Cathy Paulson, Tulsa Street Elementary School; Dawn Kelly, Hale Middle School; Christine Peña, Hollenbeck Middle School; Catherine Choi, Belmont High School; and R. Natasha Galvez, Northridge Academy High School.

The original authors of *Art & Science* were Elissa Ennis, Ben Garcia, and Terah Gonzalez. The curriculum was revised in 2013 by Theresa Sotto, education specialist for teacher audiences, and Sandy Rodriguez, project specialist.

Additional thanks to Elizabeth Escamilla, senior manager; Toby Tannenbaum, assistant director for education; and Gregory A. Dobie for supporting and advising on this project.

Introduction

The Education Department at the J. Paul Getty Museum offers a broad range of resources and programs to engage, teach, and inspire people of all ages. It supports the Getty Museum's mission to encourage the appreciation and understanding of art and its history, context, and meaning by creating learning experiences centered on works of art. The *Art & Science* curriculum was developed by Getty educators with museum conservators, curators, and scientists and a teacher advisory group. Focusing on the science of art production and conservation and the scientific skills of investigation and experimentation, it supports student proficiency in science and visual art by exploring the fascinating territory in which the arts and sciences mingle.

The lessons and discussion ideas are related to objects from the Getty Museum's collection areas: antiquities, decorative arts, drawings, manuscripts, painting, photography, and sculpture.

Science and art teachers are encouraged to collaborate when implementing these lessons to share resources and expertise. Lessons are divided into beginning, intermediate, and advanced levels that act as scaffolds for step-by-step learning. While these lessons correspond roughly to expectations for science and art learning in elementary, middle, and high schools, teachers can decide which level to use depending on their students' needs and abilities. Middle- and high-school teachers can use less advanced activities to review student understanding of basic principles required for the more complex activities.

In addition to lessons, this curriculum also contains the following resources:

- **Color Reproductions of Featured Works of Art** can be projected in the classroom with a digital document reader. The images can also be accessed online using the web addresses provided.
- **Featured Work of Art** sections provide background information on artworks and artists and include **Questions for Teaching**, which can be used to prompt examination and discussion of the featured objects.
- The **Timeline** places the featured objects within a broader historical context.
- The **Glossary** includes terms used throughout the curriculum (highlighted in **bold**).
- A **Resources** list provides starting points for deeper explorations of topics.

The *Art & Science* curriculum connects to national and California state standards. Charts for related content standards are available on the Getty website, www.getty.edu/education/teachers/classroom_resources/curricula/art_science/. Selected lessons from this curriculum, along with others related to the Getty Museum's collection, can be found online at www.getty.edu/education.

Featured Works of Art

The color reproductions of works of art can be projected in the classroom with a digital document reader. They can also be found as high-resolution digital images with zoom features on the Getty website, www.getty.edu.

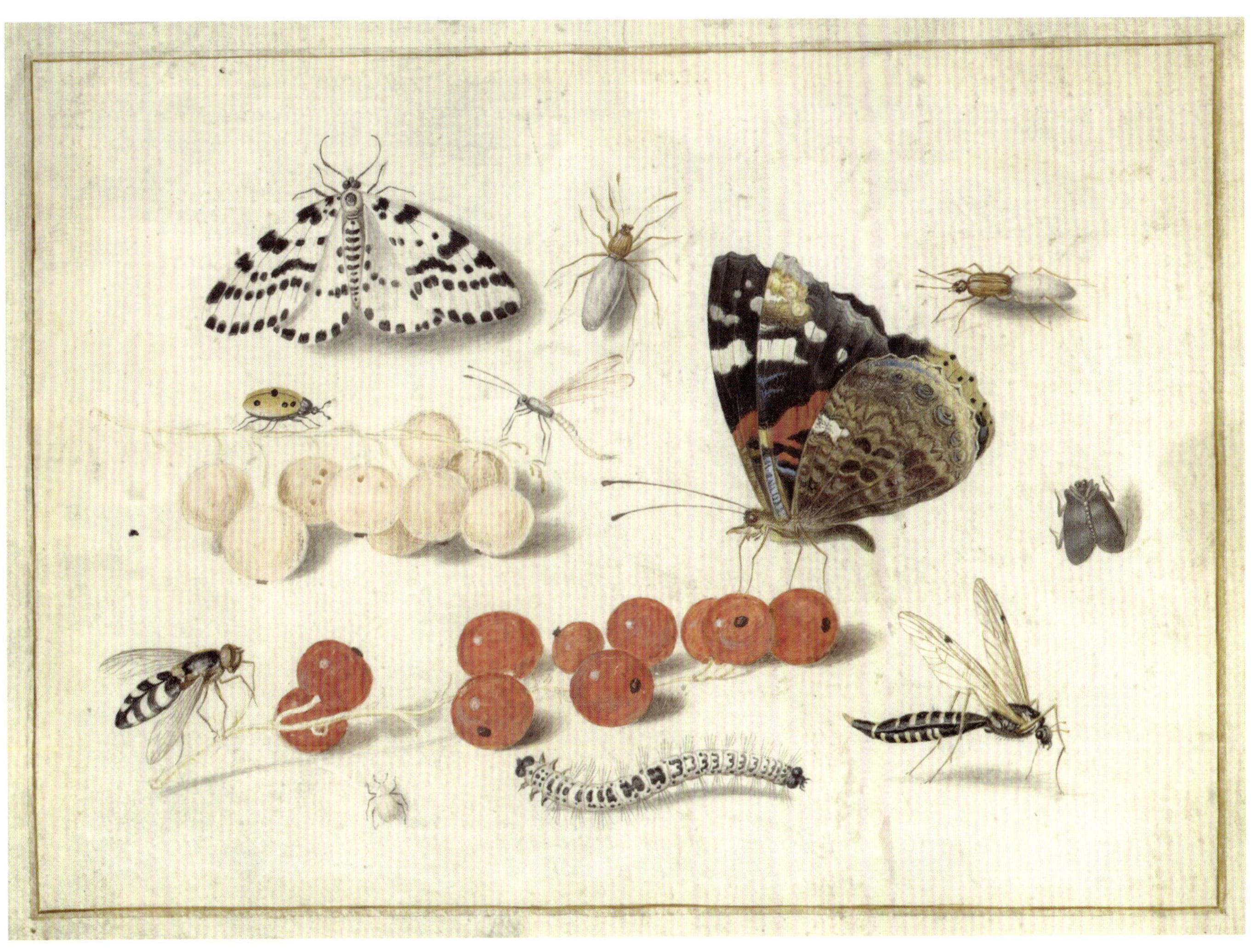

Jan van Kessel
Flemish, 1626–1679
Butterfly, Caterpillar, Moth, Insects, and Currants, 1650–55
Bodycolor and brown ink over metalpoint underdrawing on vellum
5⅛ x 6 11/16 in.
J. Paul Getty Museum
92.GC.50
www.getty.edu/art/gettyguide/artObjectDetails?artobj=393

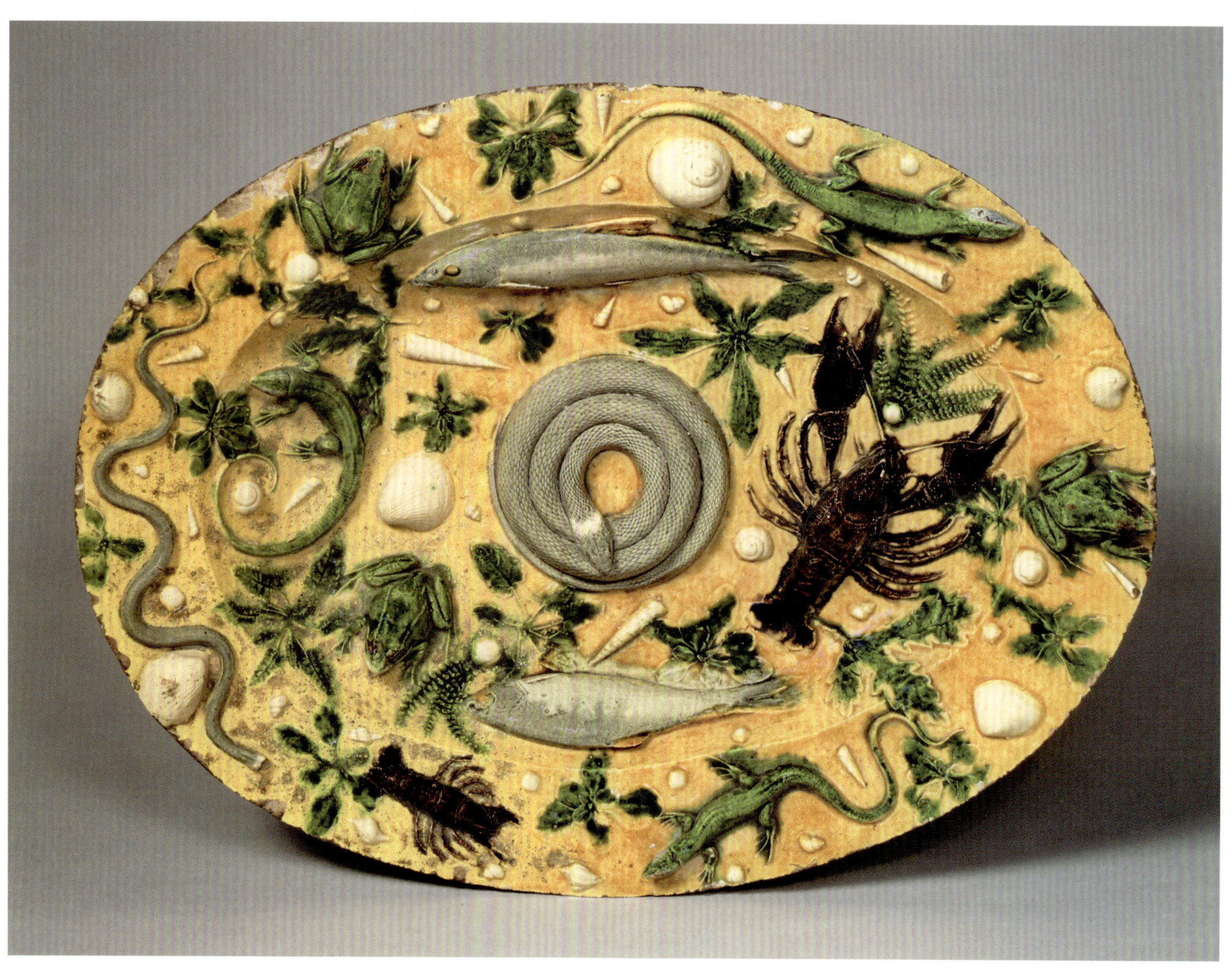

Attributed to Bernard Palissy
French, 1509–1590
Oval Basin, about 1550
Lead-glazed earthenware
18⅞ x 14½ in.
J. Paul Getty Museum
88.DE.63
www.getty.edu/art/gettyguide/artObjectDetails?artobj=1399

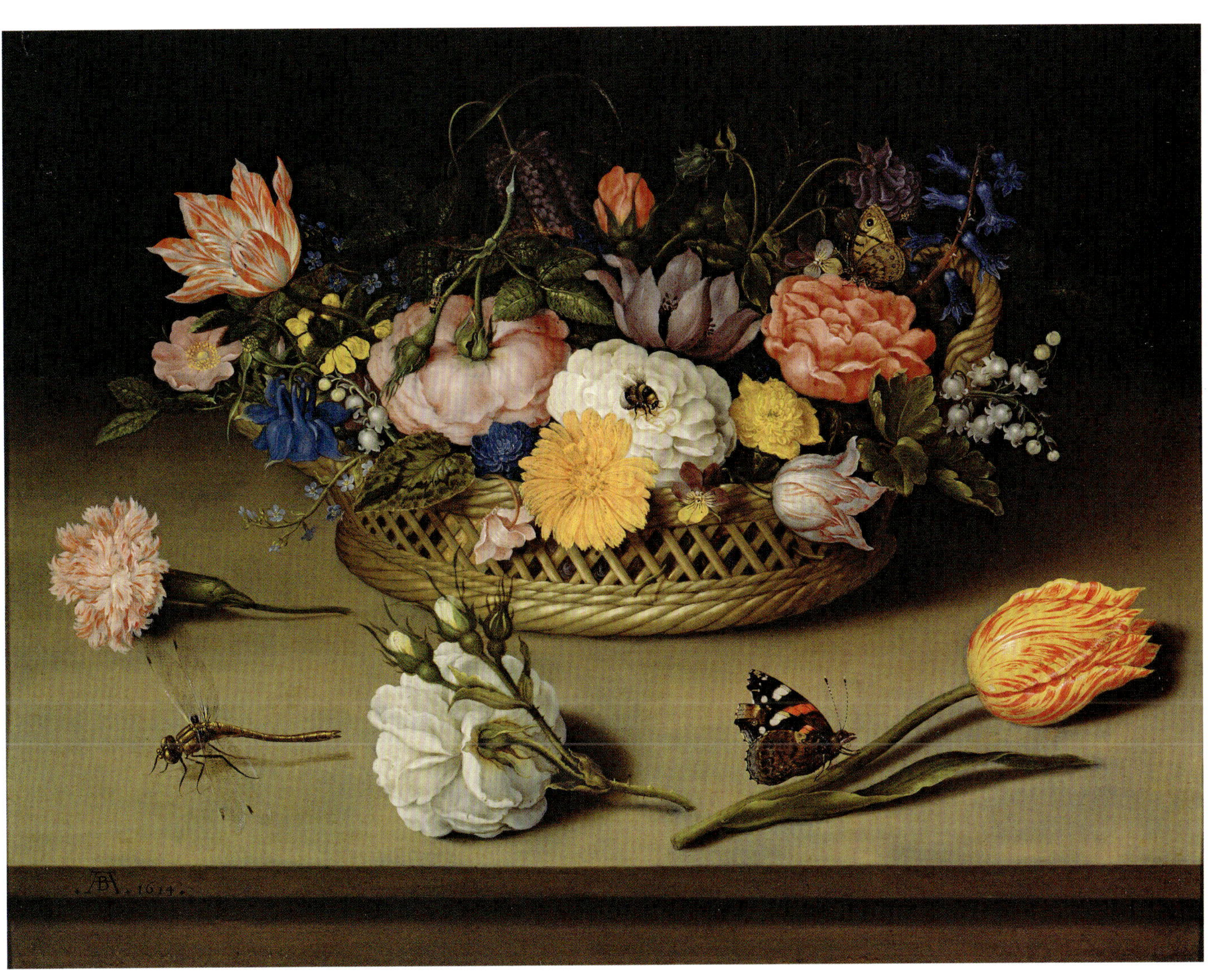

Ambrosius Bosschaert the Elder
Dutch, 1573–1621
Flower Still Life, 1614
Oil on copper
11¼ x 15 in.
J. Paul Getty Museum
83.PC.386
www.getty.edu/art/gettyguide/artObjectDetails?artobj=842

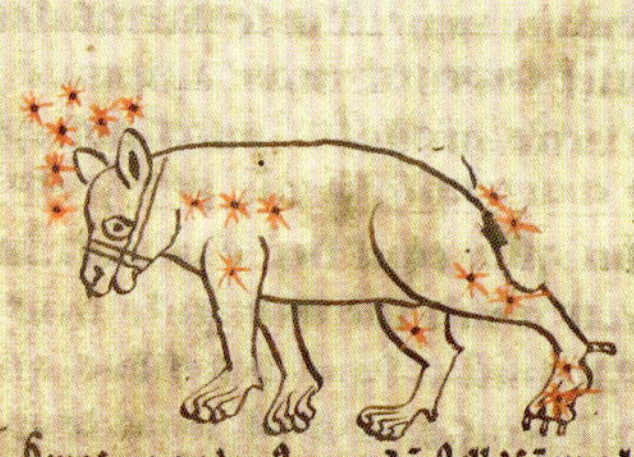

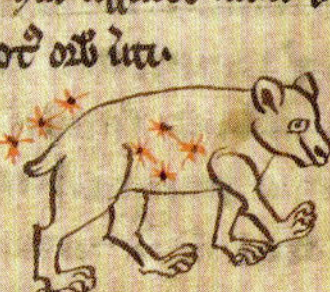

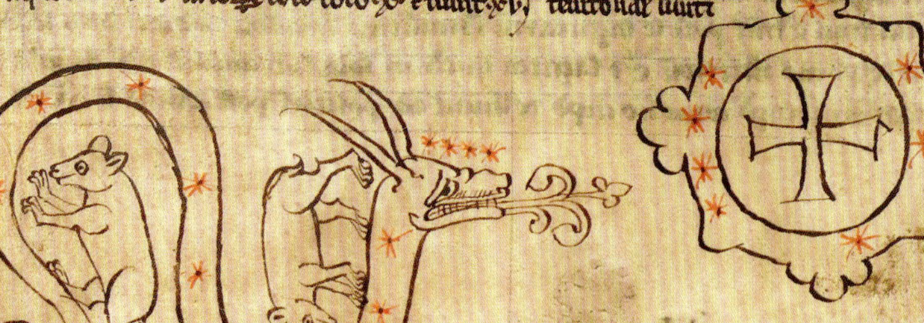

Astrological Cycle with Commentary
From the manuscript *Astronomical Miscellany*
English, early 1200s
Black, green, and red inks on parchment
9½ x 6⅛ in.
J. Paul Getty Museum
Ms. Ludwig XII 5, fol. 149v
www.getty.edu/art/gettyguide/artObjectDetails?artobj=136048

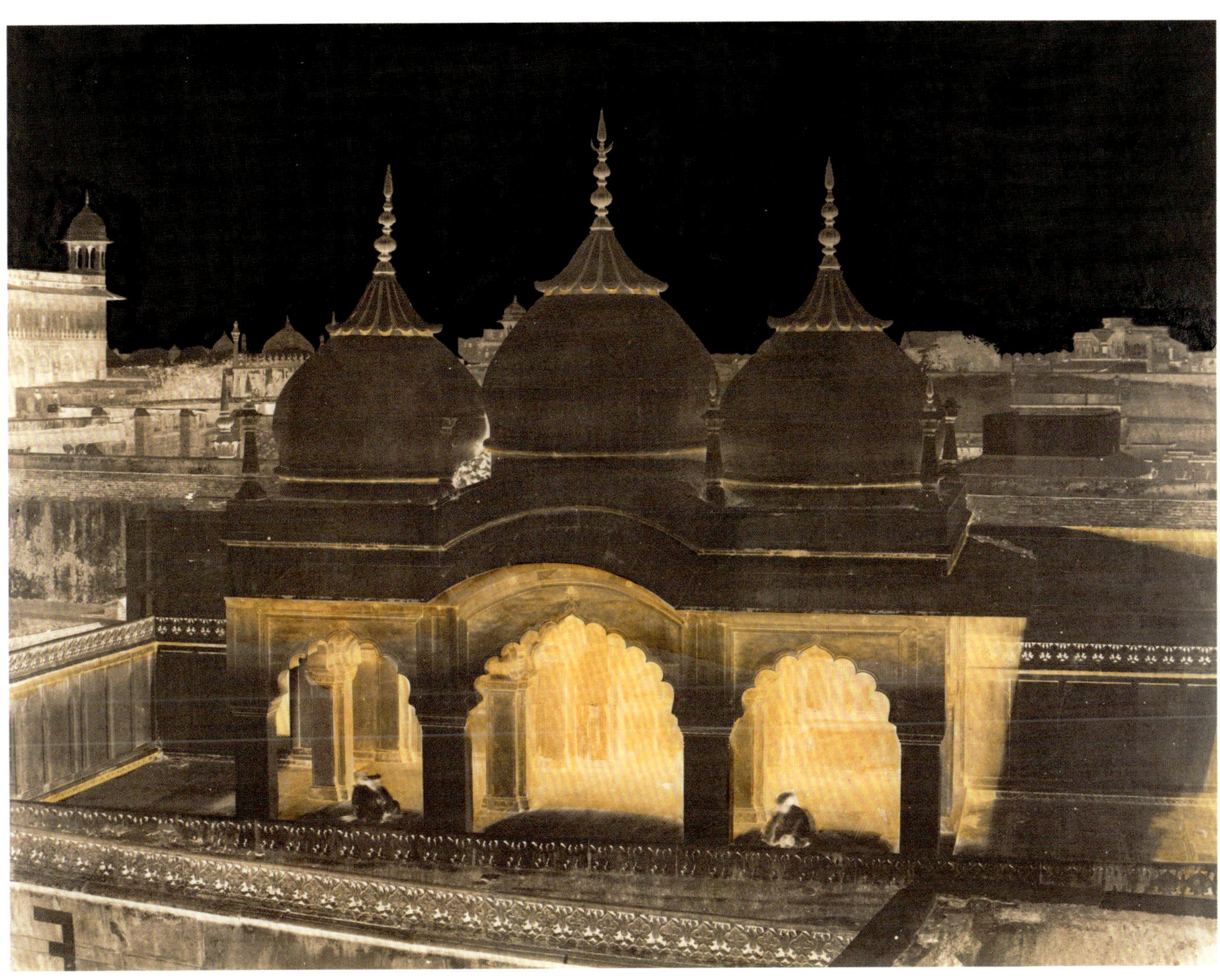

Dr. John Murray
Scottish, 1809–1898
The Emperor's Private Mosque in the Marble Palace, Agra Fort, India, 1857–60
Waxed-paper negative
14½ x 18¹⁄₁₆ in.
J. Paul Getty Museum
98.XM.7
www.getty.edu/art/gettyguide/artObjectDetails?artobj=114911

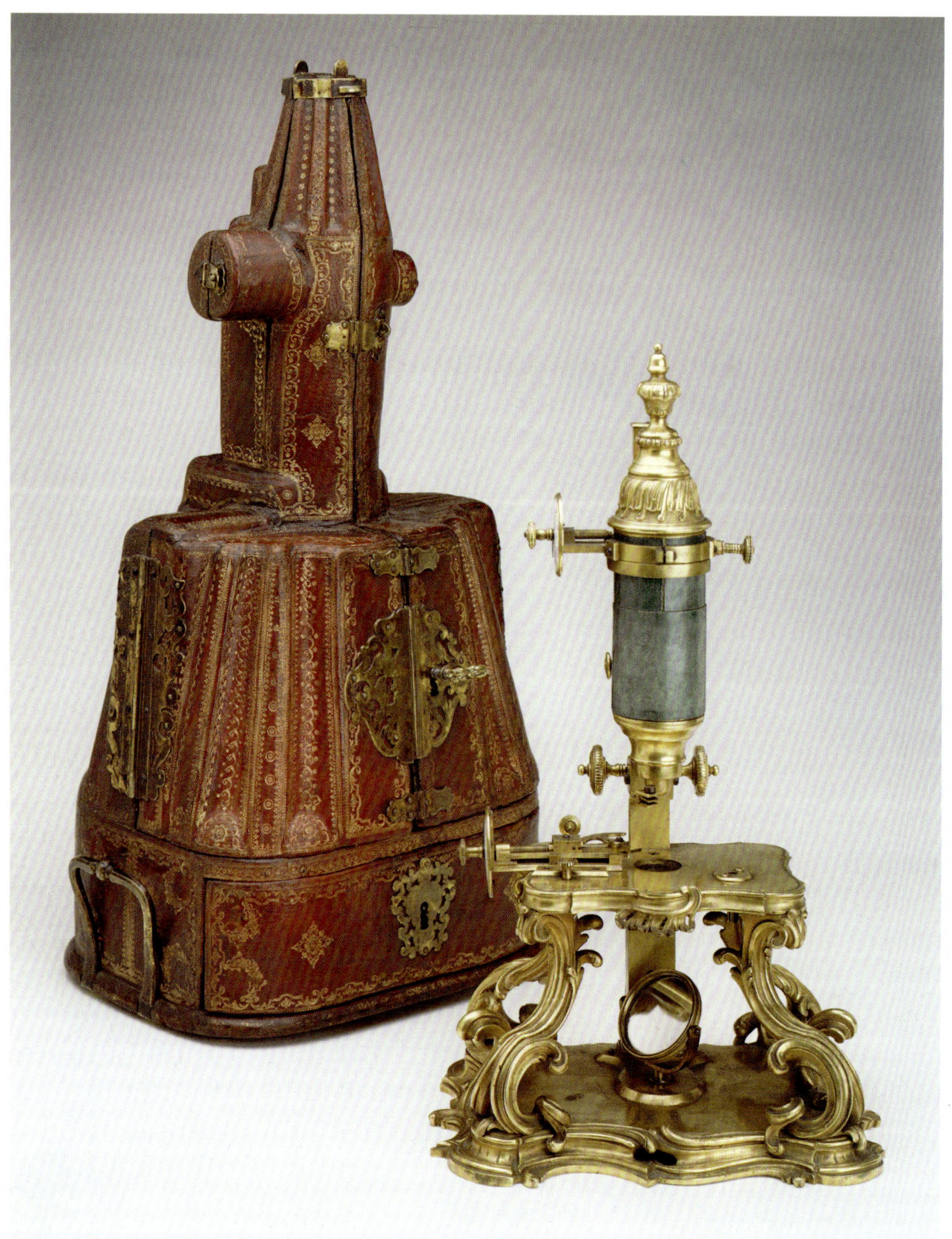

Compound Microscope and Case, about 1751
Gilt bronze attributed to Jacques Caffieri
French, 1678–1755
Micrometric stage invented by Michel-Ferdinand d'Albert d'Ailly, duc de Chaulnes
French, 1714–1769
MICROSCOPE: gilt bronze, enamel, shagreen, and glass
18⅞ x 11 x 8¹⁄₁₆ in.
CASE: wood, gilded leather, brass, velvet, silver braid, and silver lace
26 x 13¾ x 10⅝ in.
J. Paul Getty Museum
86.DH.694
www.getty.edu/art/gettyguide/artObjectDetails?artobj=6789

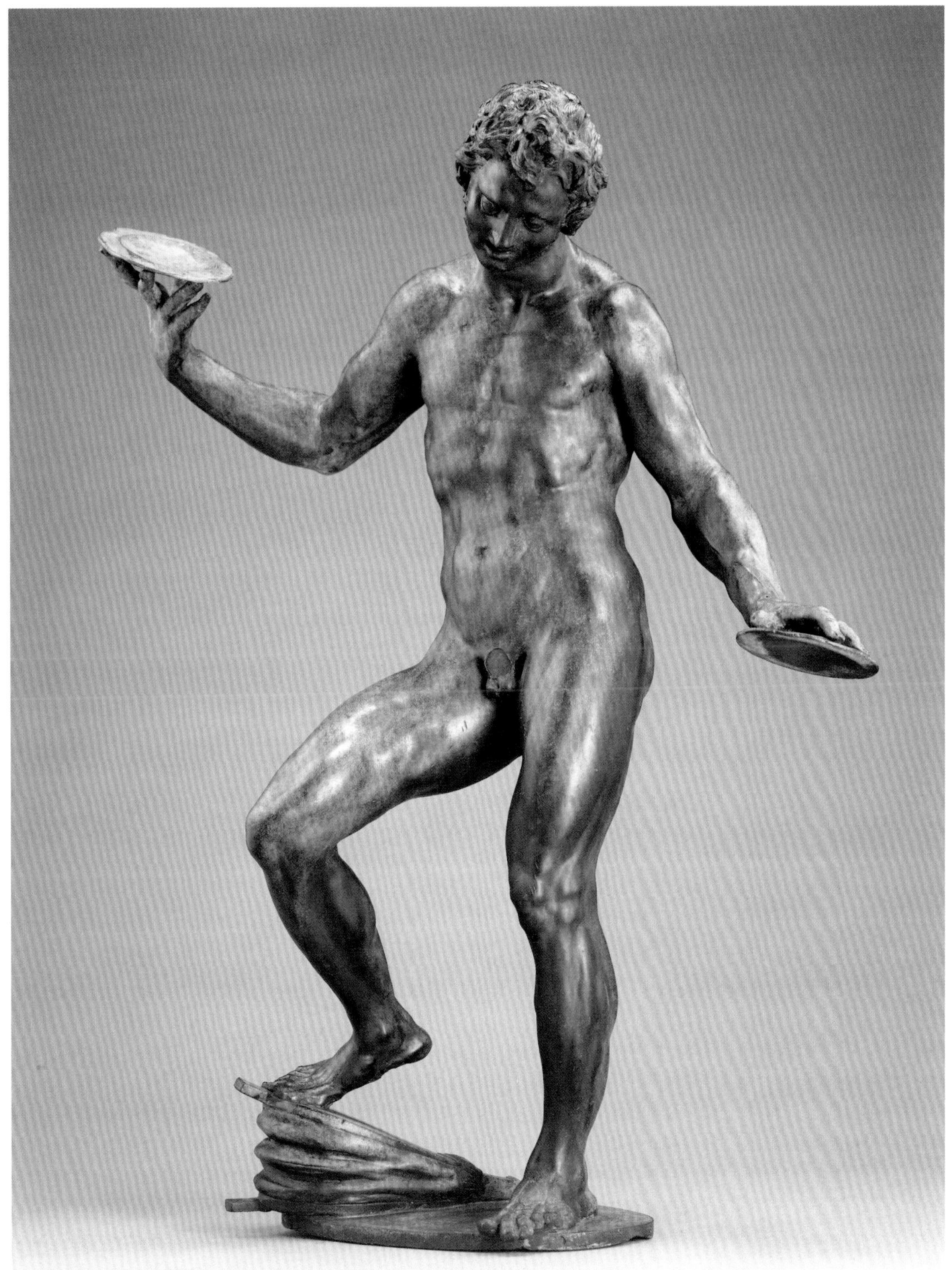

Adriaen de Vries
Dutch, about 1545–1626
Juggling Man, 1610–15
Bronze
30¼ x 20⅜ x 8⅝ in.
J. Paul Getty Museum
90.SB.44
www.getty.edu/art/gettyguide/artObjectDetails?artobj=1410

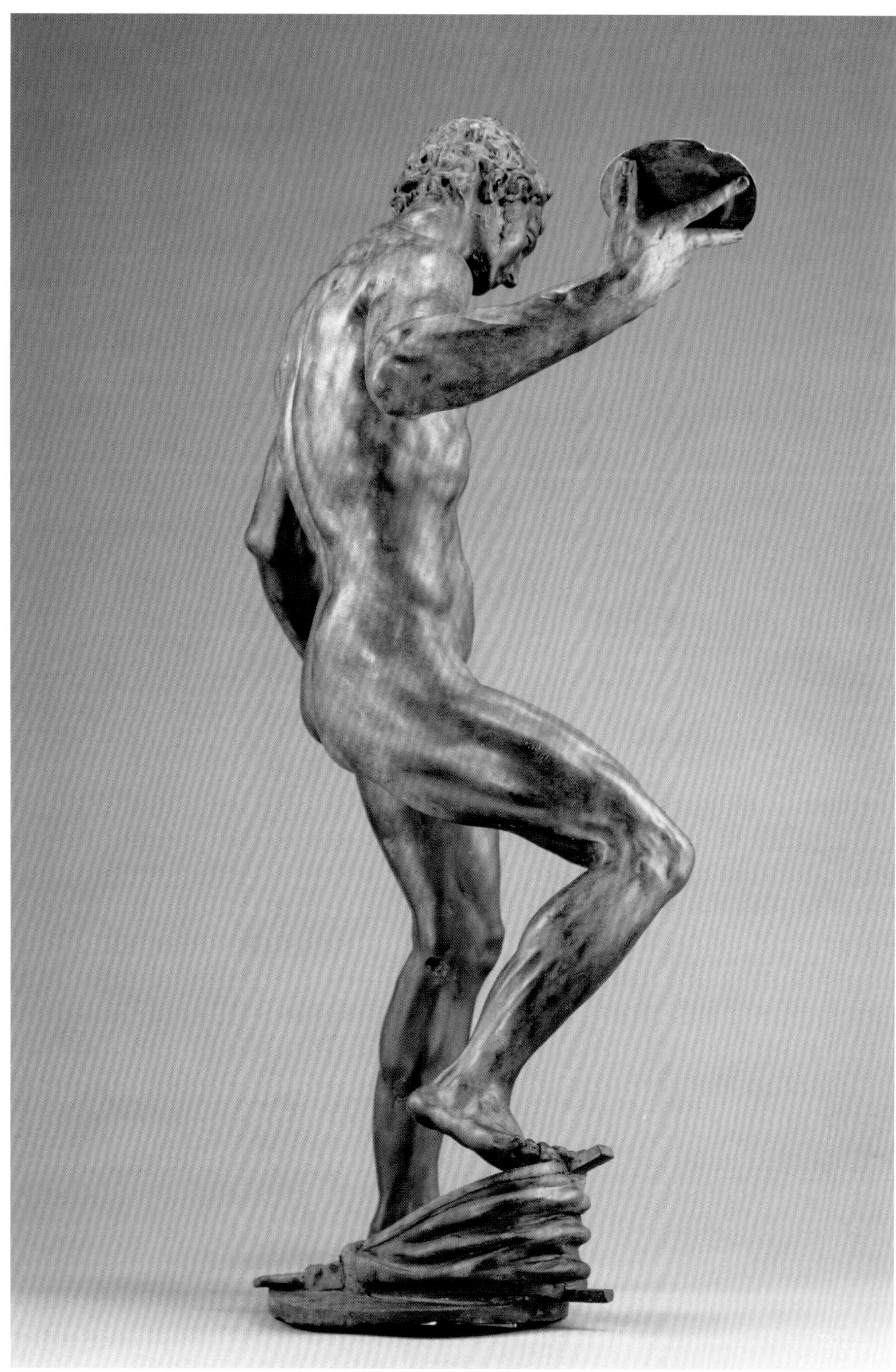

Adriaen de Vries
Dutch, about 1545–1626
Juggling Man, 1610–15
Bronze
$30\frac{1}{4}$ x $20\frac{3}{8}$ x $8\frac{5}{8}$ in.
J. Paul Getty Museum
90.SB.44
www.getty.edu/art/gettyguide/artObjectDetails?artobj=1410

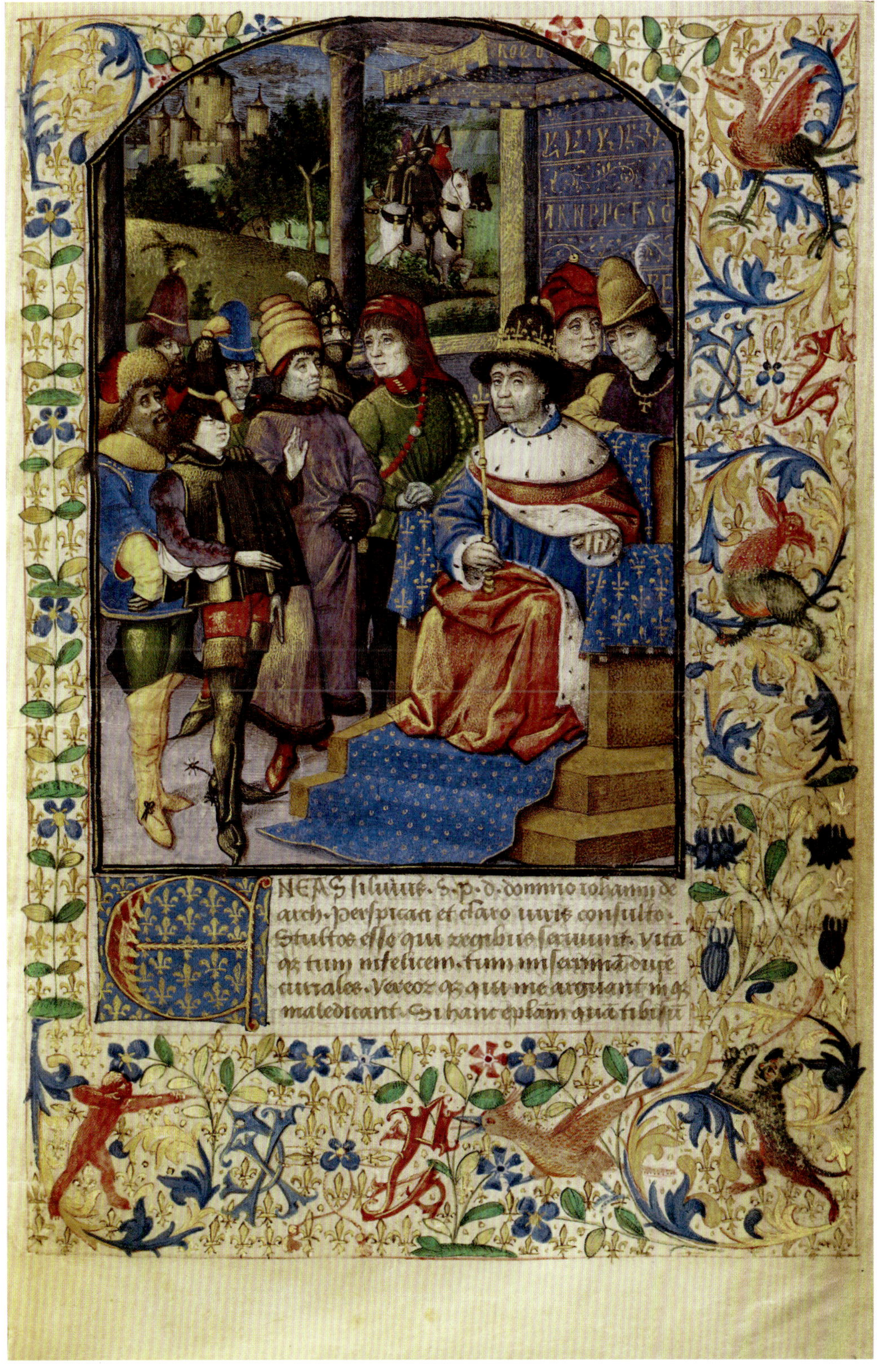

The French King at Court
From the manuscript *The Story of Two Lovers*
French, 1460–70
Tempera colors and gold paint on parchment
6 15/16 x 4½ in.
J. Paul Getty Museum
Ms. 68, fol. 1
www.getty.edu/art/gettyguide/artObjectDetails?artobj=143858

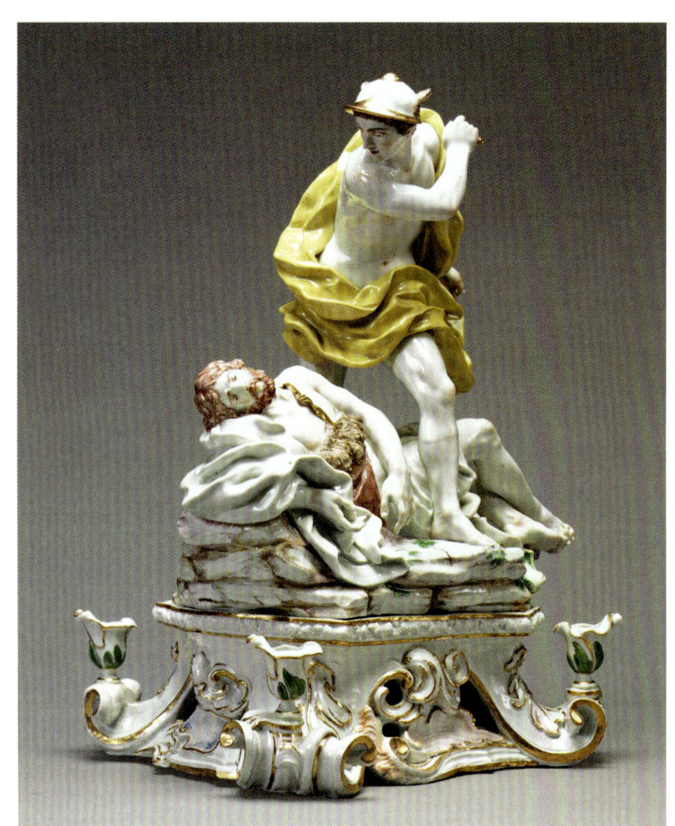

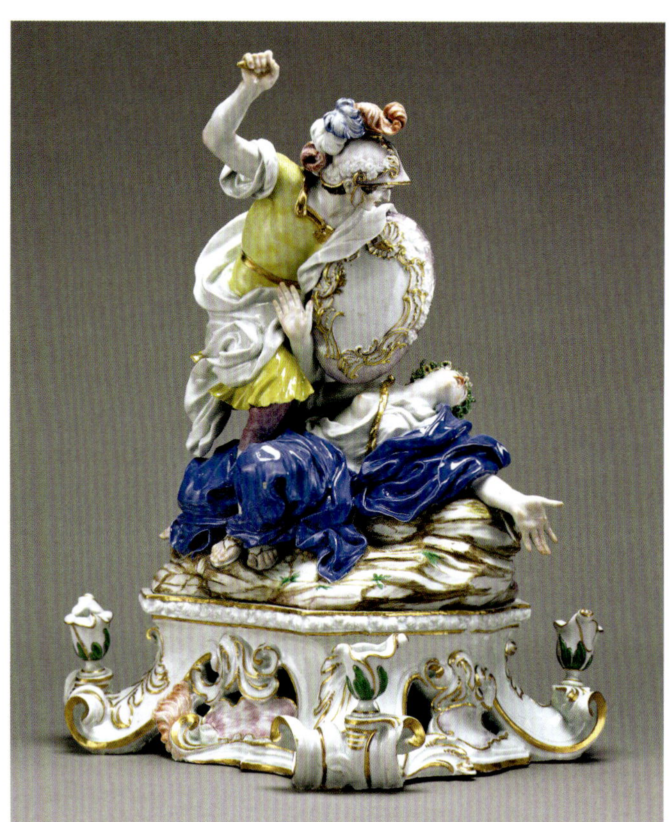

Ginori Porcelain Factory
Italian, 1735–present
Mercury and Argus and *Perseus and Medusa*, 1749
After models by Giovanni Battista Foggini
Italian, 1652–1725
Porcelain, polychrome enamel, and parcel gilt
Each 17¾ x 13 x 11 in.
J. Paul Getty Museum
94.SE.76
www.getty.edu/art/gettyguide/artObjectDetails?artobj=1478

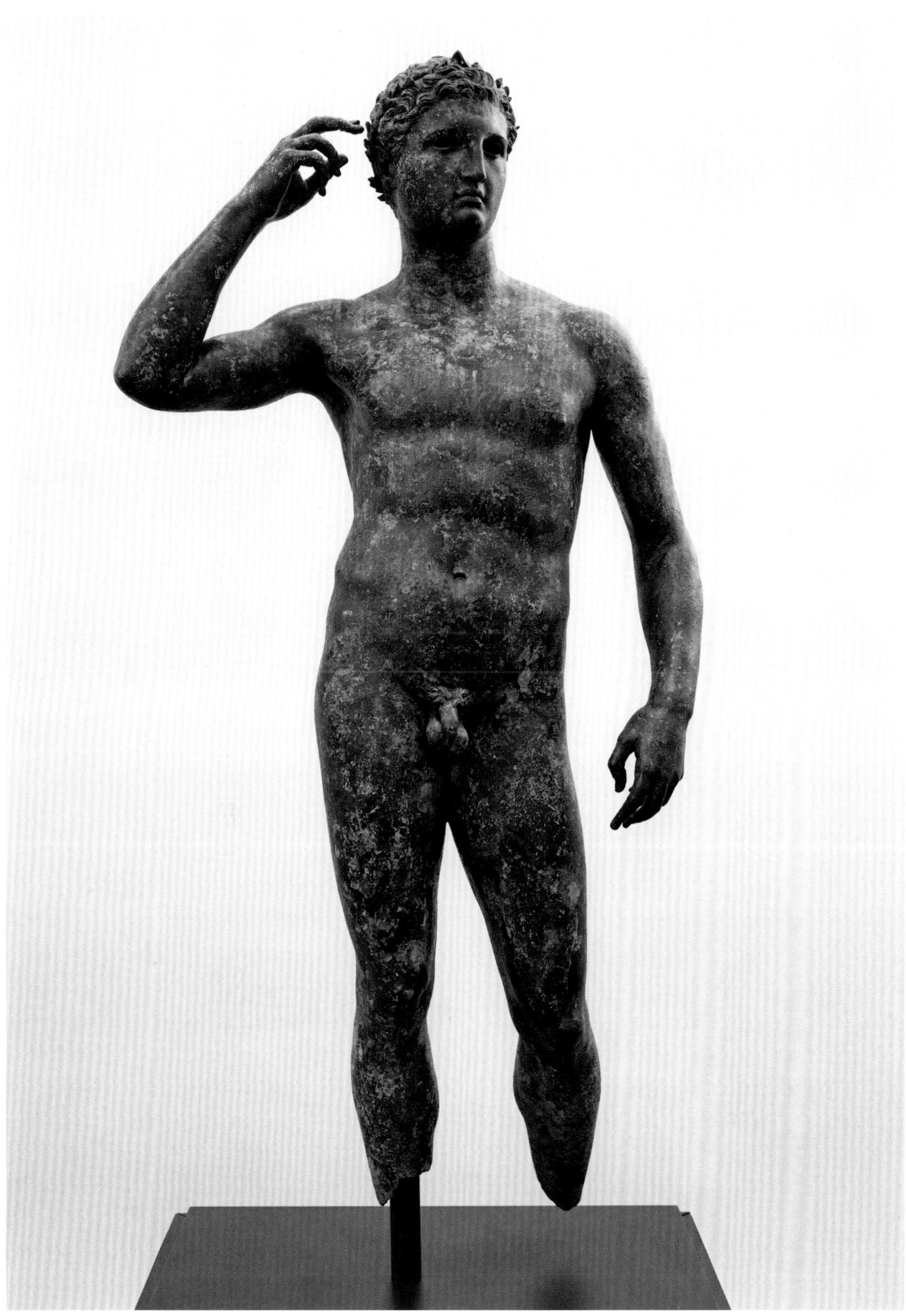

Victorious Youth
Greek, 300–100 B.C.
Bronze
59 5/8 x 27 9/16 x 11 in.
J. Paul Getty Museum
77.AB.30
www.getty.edu/art/gettyguide/artObjectDetails?artobj=891

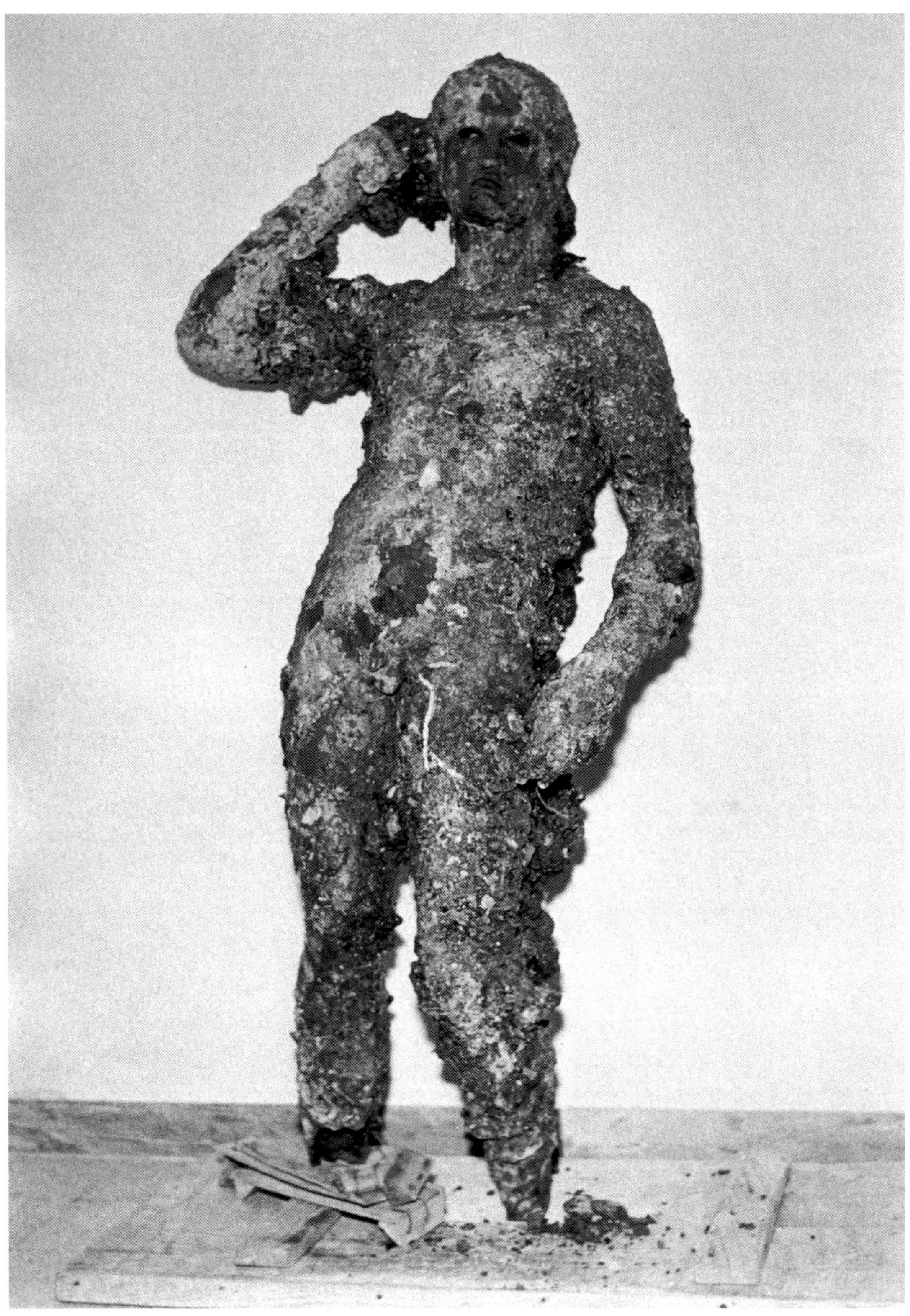

Victorious Youth
Greek, 300–100 B.C.
Bronze
59⅝ x 27⁹⁄₁₆ x 11 in.
J. Paul Getty Museum
77.AB.30
www.getty.edu/art/gettyguide/artObjectDetails?artobj=891

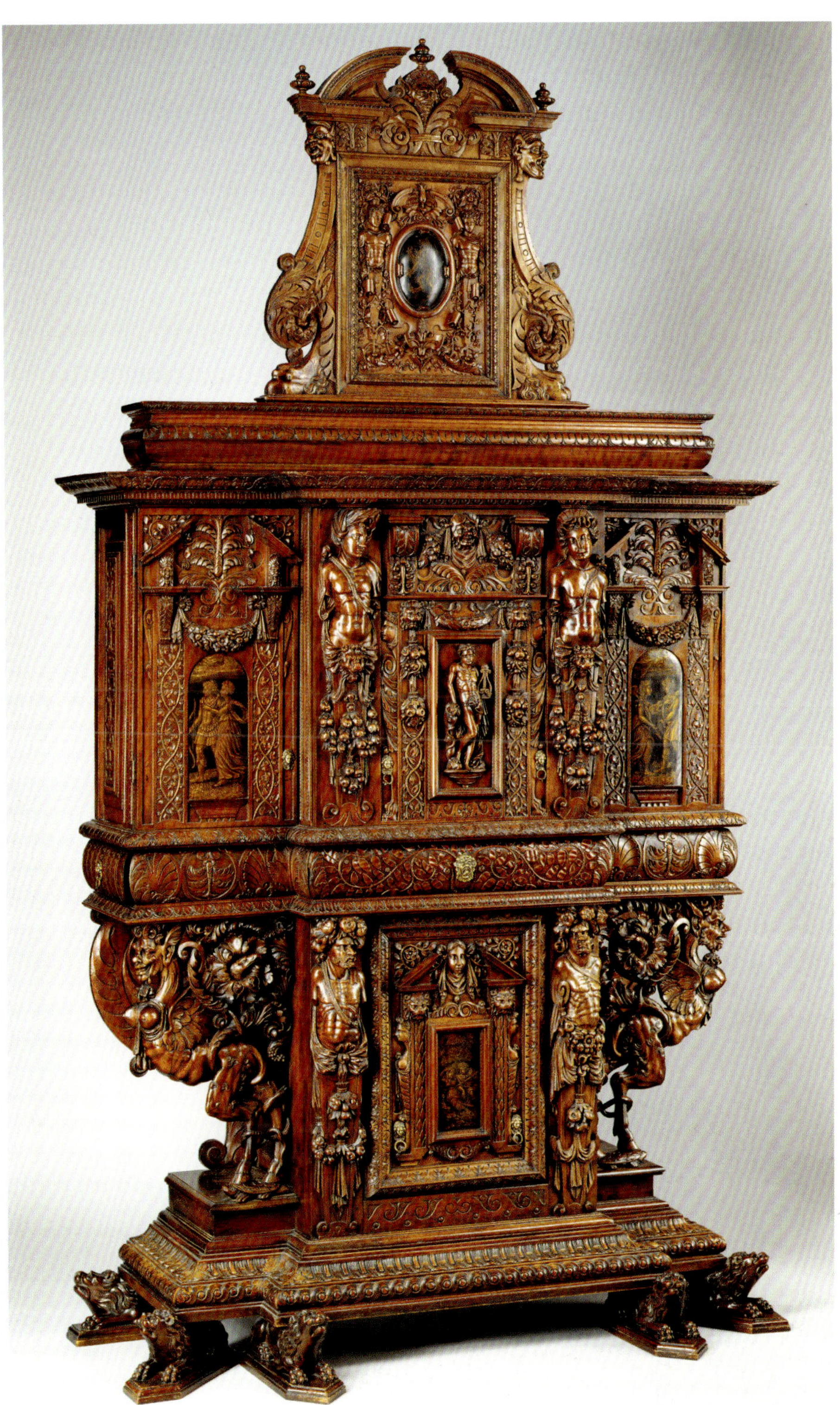

Cabinet
French, 1580
Carved walnut and oak with painted panels, linen and silk lining
121⅛ x 60⅜ x 22½ in.
J. Paul Getty Museum
71.DA.89
www.getty.edu/art/gettyguide/artObjectDetails?artobj=1104

Insect Anatomy and the Scientist as Illustrator

Using a seventeenth-century entomological drawing, students study and identify a variety of insects. They identify some characteristics common to all insects and others that are unique to particular species. They engage their observational skills to create detailed drawings of insects.

Beginning Level
Students focus on the three characteristics of an adult insect: a three-part body (head, thorax, and abdomen), six legs, and antennae. Students collect and draw live insects, incorporating a variety of shapes and lines.

Intermediate Level
Students research an insect depicted in a seventeenth-century drawing and then draw it, incorporating a variety of lines and shapes. They explore how value is used to create the illusion of three-dimensionality in a drawing.

Advanced Level
Students study and research winged insects, identifying unique characteristics and those common to all insects. They closely observe and sketch a winged insect and create a detailed drawing of wings.

Standards Addressed
Refer to the charts for national and California state standards on the Getty website, www.getty.edu/education/teachers/classroom_resources/curricula/art_science/downloads/standards.pdf

FEATURED WORK OF ART ►

Jan van Kessel
Flemish, 1626–1679
Butterfly, Caterpillar, Moth, Insects, and Currants, 1650–55
Bodycolor and brown ink over metalpoint underdrawing on vellum
5⅛ x 6¹¹⁄₁₆ in.
J. Paul Getty Museum
92.GC.50
www.getty.edu/art/gettyguide/artObjectDetails?artobj=393

In this drawing, luminous red and white currants provide perches for various **insects**, while others, including a moth, ladybug, caterpillar, and fly, are set in different places around the page. Jan van Kessel was one of the first artists to create horizontally arranged drawings of fruits and flowers, giving structure to scientific analyses of natural **specimens**. He used cast shadows to suggest a pictorial space in his drawings. Instead of a specific setting, he preferred to use a blank ground, which allowed him to arrange the insects and fruits according to shifting yet interconnected vantage points. Van Kessel probably made this drawing in the studio, based on close sketches from nature. He used **bodycolor** to show volume, as seen in the luminous, patterned wings of the moth and butterfly. Transparent **washes** create **light** effects; a light source at the top left creates dark shadows along the right sides of wings and underneath the insects' bodies, lifting them from the ground. Van Kessel assembled these insects to emphasize each specimen's distinct forms and markings.

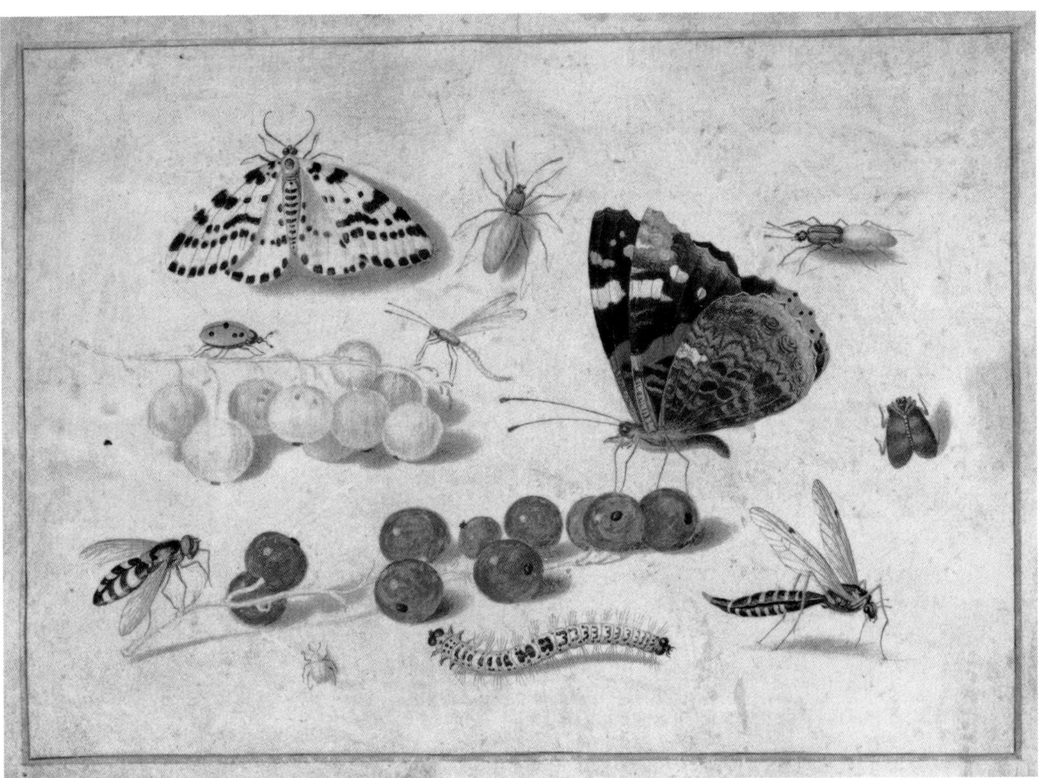

ABOUT THE ARTIST

Jan van Kessel counted his uncle **Jan Brueghel the Younger** among his teachers. He joined the Antwerp painters' guild in 1645 and specialized in small-scale pictures of subjects taken from the natural world, such as floral still lifes and **allegorical** series showing animal kingdoms, the four elements, the senses, or the parts of the world. Obsessed with picturesque detail, Van Kessel worked from nature and used **scientific illustrations** as sources for the subjects of his pictures.

Scholars trace many of Van Kessel's subjects back to prototypes by his predecessors. **Joris Hoefnagel**'s works inspired Van Kessel's sensitive and delicate drawings of insects and flowers, executed mainly in watercolor on parchment. Van Kessel showed a preference for beetles, caterpillars, and butterflies and occasionally arranged caterpillars to spell out his name. The works of **Roelandt Savery**, **Frans Snyders**, and Van Kessel's grandfather **Jan Brueghel the Elder** influenced his paintings of animals.

QUESTIONS FOR TEACHING

- What do you see in this drawing?
- What types of lines do you see?
- What types of shapes? Which shapes are organic and which are geometric?
- Where do you see similar lines or shapes?
- Which objects or insects are similar to each other?
- How are they similar? What do these similarities tell you about how these things are related to each other?
- Notice the negative space around the objects. How does this affect the way you look at the insects and fruit?
- How do these insects compare to those you might find in nature?
- This image relates to the tradition of collecting specimens and pinning them to a blank surface for observation. How does Van Kessel go beyond strict observation with this drawing?

LESSON PLAN | BEGINNING LEVEL

Grades: Lower elementary (K–2), upper elementary (3–5)
Subjects: Science and visual arts
Time required: 2–3 class periods

Lesson Overview

Students observe live **insects** and examine insects depicted in a seventeenth-century drawing. They identify the three characteristics of an adult insect: a three-part body (**head**, **thorax**, and **abdomen**), six legs, and **antennae**. They collect and draw live insects, incorporating a variety of **shapes** and **lines**.

Learning Objectives

Students will:

- Use observation and prior knowledge to describe characteristics shared by all insects.
- Describe and identify insects in a seventeenth-century drawing.
- Draw an insect from close observation, using a variety of shapes and lines.

Materials

- Reproduction of *Butterfly, Caterpillar, Moth, Insects, and Currants* (p. 7)
- Information about the featured work of art and Questions for Teaching (pp. 20–21)
- Clear containers with lids with small pinholes poked in the top
- Paper, pencils, and colored pencils
- Art activity: "Drawing Insects with Organic Shapes and Lines," www.getty.edu/education/teachers/classroom_resources/tips_tools/downloads/drawing_insects.pdf

Lesson Steps

1. Display *Butterfly, Caterpillar, Moth, Insects, and Currants* by Jan van Kessel or hand out reproductions. Tell students to look closely at the drawing and share their initial observations. Then ask students the following:
   - What do you see in this drawing?
   - What types of lines do you see?
   - What types of shapes?
   - Where do you see similar lines or shapes?
   - Which objects or insects are similar to each other?

2. Inform students that before **photography**, scientists had to rely on drawings of natural **specimens** to study insects, and artists were often called upon to make these drawings. **Entomological** still-life painting became an important **genre** in Europe. Tell students to use their prior knowledge to identify the different insects they see in the drawing. Chart the names of the different insects that can be seen in the drawing.

3. Ask students for a definition of an insect and chart the responses. Select from this list the three universal characteristics of an adult insect: a three-part body (head, thorax, and abdomen), six legs, and antennae.

4. Return to the drawing and ask, "Which of these bugs do not have the three characteristics of adult insects?" Focus on the caterpillar and ask students what they know about caterpillars. Chart responses. Explain that caterpillars are larvae for butterflies and moths. Have students identify which of the insects were once caterpillars (moth and butterfly). Explain that some insects have wings that cover their bodies (such as the fly and ladybug), but that they have the same three parts to their bodies as the other insects.

5. Have students collect insects from the schoolyard in clear containers and cover them with lids that have been poked with pinholes. Hand out paper and art supplies. Have students draw their insects, making sure to identify the three characteristics outlined above. Remind students of the lines and shapes they found when discussing the drawing by Jan van Kessel (e.g., ovals for the head and thorax, thin lines for the legs). Encourage students to draw their insects using a variety of lines and shapes. You may wish to view the art activity "Drawing Insects with Organic Shapes and Lines" on the Getty website.

6. Release the insects back to the area in which they were found.

LESSON PLAN | INTERMEDIATE LEVEL

Grades: Middle school (6–8)
Subjects: Science and visual arts
Time required: 3 class periods

Lesson Overview

Students observe and study **insects** depicted in a seventeenth-century drawing. They identify characteristics common to all insects and those unique to particular **species**. Students research and draw insects, incorporating a variety of lines and shapes and using **value** to depict three-dimensionality.

Learning Objectives

Students will:

- Use observation, research, and prior knowledge to describe the universal and unique characteristics of insects.
- Describe and identify insects in a seventeenth-century drawing.
- Draw insects using a variety of shapes and lines.
- Use value to create the illusion of three-dimensionality in their drawings.

Materials

- Reproduction of *Butterfly, Caterpillar, Moth, Insects, and Currants* by Jan van Kessel (p. 7)
- Four detail reproductions of insects from the painting: www.getty.edu/education/teachers/classroom_resources/curricula/art_science/00039301_detail1.html www.getty.edu/education/teachers/classroom_resources/curricula/art_science/00039301_detail2.html www.getty.edu/education/teachers/classroom_resources/curricula/art_science/00039301_detail3.html www.getty.edu/education/teachers/classroom_resources/curricula/art_science/00039301_detail4.html
- Information about the featured work of art and Questions for Teaching (pp. 20–21)
- Student handout: Insect Fact Sheet (p. 24)
- Drawing paper
- Pencils and colored pencils
- Art activity, "Drawing Insects with Organic Shapes and Lines:" www.getty.edu/education/teachers/classroom_resources/tips_tools/downloads/drawing_insects.pdf
- Art activity: "Exploring Value to Create Form," www.getty.edu/education/teachers/classroom_resources/tips_tools/downloads/exploring_value.pdf

Lesson Steps

1. Complete steps 1–4 of the beginning-level lesson, adapting for grade level as appropriate. For example, when viewing the drawing by Van Kessel, you may wish to discuss **color** and **composition** in addition to shapes and lines.

2. Divide the class into pairs and hand out pages with details of the insects in Van Kessel's drawing, so that each pair of students has an assigned insect.

3. Give students access to the Internet or science texts and ask them to research their assigned insect. You can also prepare information sheets about all of the insects for students to use as research material.

4. Hand out paper and art supplies. Have each student draw his or her insect, making sure to identify the **head**, **thorax**, **abdomen**, legs, **antennae**, and **mandibles** (when relevant). Remind students of the lines and shapes they identified when discussing the drawing by Van Kessel by pointing out examples (e.g., ovals for the head and thorax, thin lines for the legs). Encourage students to draw their insects using a variety of lines and shapes. You may wish to view the art activity "Drawing Insects with Organic Shapes and Lines" on the Getty website.

5. Direct students' attention to their details again. Ask them to consider what the artist did to make the insects look three-dimensional. Point out examples of shadows and shading. Students should use value to give their insect drawings a three-dimensional appearance. You may wish to view the art activity "Exploring Value to Create Form" on the Getty website. Ask students if drawing their **specimens** helped them to observe something they did not notice before.

6. Students should label the various body parts of their insects. They should label the three characteristics that are universal to all insects and also identify those that are specific to their insect's species.

7. Have students work together to research and fill in the Insect Fact Sheet student handout. Display the students' fact sheets with their labeled insect drawings.

STUDENT HANDOUT

INSECT FACT SHEET

Use this worksheet to record your findings about the insect you are researching.

1. What is its lifespan?

2. What does it eat?

3. Does it live in groups or by itself?

4. Where does it live?

5. Does it fly?

6. What is the most interesting thing about it?

LESSON PLAN | ADVANCED LEVEL

Grades: High school (9–12)
Subjects: Science and visual arts
Time required: 3–4 class periods

Lesson Overview

- Students study **insects** depicted in a seventeenth-century drawing. They research winged insects, identifying unique characteristics and those common to all insects. Students closely observe winged insects and create detailed drawings of wings.

Learning Objectives

Students will:

- Use observation, research, and prior knowledge to describe characteristics shared by all insects.
- Use knowledge about insect anatomy to differentiate among various **species** of insects represented in a seventeenth-century drawing.
- Understand that insects are **invertebrates** with **exoskeletons**.
- Draw insects using a variety of **shapes**, **lines**, and textures.
- Use **value** to create the illusion of three-dimensionality in drawings.

Materials

- Materials listed in the intermediate-level lesson (p. 23)
- Winged-insect specimens
- Tweezers
- Scalpels
- Magnifying glasses

Lesson Steps

1. Complete steps 1–4 of the beginning-level lesson, adapting for grade level as appropriate.

2. Complete steps 2–7 of the intermediate-level lesson, adapting the steps to focus only on the winged insects depicted in Jan van Kessel's drawing.

3. Introduce the terms *invertebrate* and ***vertebrate*** to students. List types of animals that fall into each category.

4. Once you have identified insects as invertebrates, ask students to speculate about how insects move and keep their shapes without internal bones. Discuss exoskeletons and their functions. Explain that wings are part of the exoskeleton.

5. Hand out **specimens** of houseflies, butterflies, dragonflies, moths, wasps, or any other winged insect that is represented in Van Kessel's drawing.

6. Have students use tweezers and scalpels to detach the insects' wings from the points at which they join the body. Using magnifying glasses and colored pencils, students will draw diagrams of wings, including as much detail as possible. Students should use line, shape, and **color** to visually describe the various textures of the wings. On the same page as the drawing, have students write a paragraph describing the wings to someone who cannot see them.

7. Instruct students to work with a partner to compare and contrast their drawings of insect wings with the wings depicted in Van Kessel's drawing. Students should speculate about how various characteristics of wings could allow different insects to move in distinct ways. Have student pairs share their speculations with the entire class.

Cast from Life
A Marshland Ecosystem in a Bowl

Students classify the plants and animals depicted on a sixteenth-century ceramic basin and determine what ecosystem is represented in the artwork. They study how balance is created by the arrangement of plants and animals on the basin and create their own decorative art objects inspired by the original sixteenth-century work.

Beginning Level
Students classify the plants and animals depicted on the basin as herbivores, carnivores, or omnivores. They research an ecosystem to depict on their own visually balanced decorative plates.

Intermediate Level
Students classify the plants and animals depicted on the basin based on diet and whether they are producers, consumers, or decomposers. Students create food webs based on both the sixteenth-century basin and their own decorative plates. Their plates will depict living organisms in a selected ecosystem and demonstrate their understanding of visual balance.

Advanced Level
Students expand on the intermediate activities, focusing on understanding and depicting balanced ecosystems.

Standards Addressed
Refer to the charts for national and California state standards on the Getty website, www.getty.edu/education/teachers/classroom_resources/curricula/art_science/downloads/standards.pdf

FEATURED WORK OF ART ►

Attributed to Bernard Palissy
French, 1509–1590
Oval Basin, about 1550
Lead-glazed earthenware
18⅞ x 14½ in.
J. Paul Getty Museum
88.DE.63
www.getty.edu/art/gettyguide/artObjectDetails?artobj=1399

A man of many interests and talents, Bernard Palissy developed his distinctive "rustic-ware" by **casting** shellfish, plants, and reptiles from life. He then attached the shapes to traditional **ceramic** forms, such as **ewers** and plates, and painted them in blue, green, purple, and brown. Finally, he decorated the wares with runny, lead-based **glazes**, which increased their watery realism.

Palissy's works were highly popular. Such influential patrons as Catherine de Medici, queen of France, commissioned him to decorate **grottoes** in their private gardens, settings for diversion and contemplation in the 1500s. Palissy's rustic works were so successful that they were imitated during his own lifetime. In the 1800s, such notable ceramic factories as Sèvres in France and Wedgwood in England also copied them.

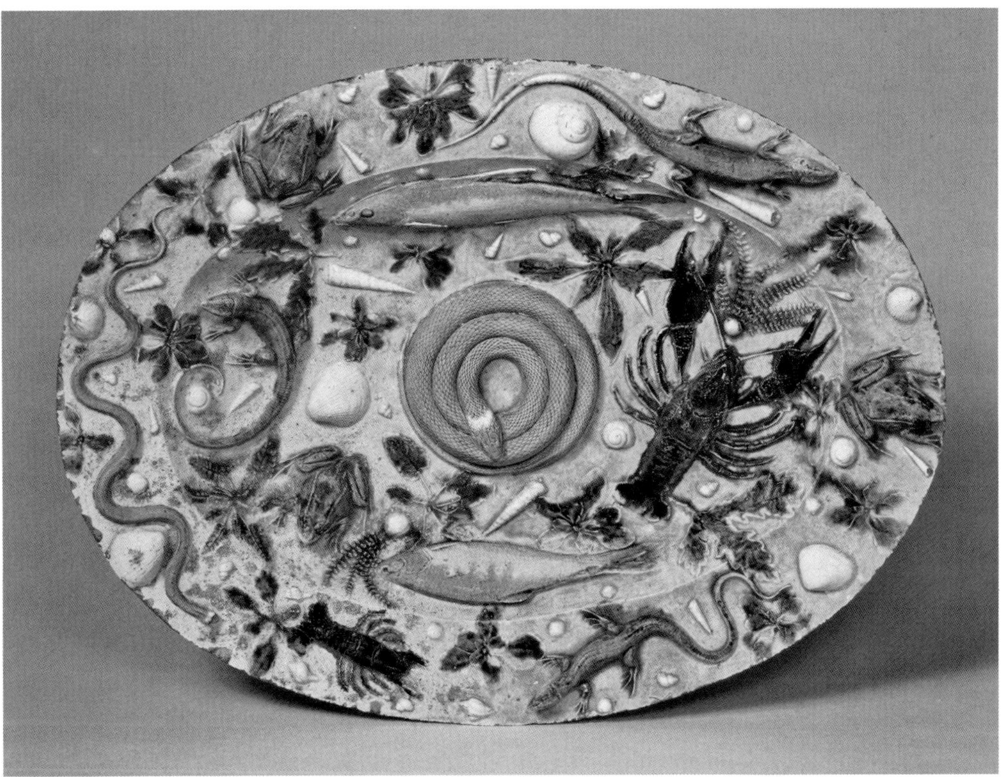

ABOUT THE ARTIST

Although having no formal training, Bernard Palissy became a scientist, surveyor, religious reformer, garden designer, glassblower, painter, chemist, geologist, philosopher, and writer, as well as a ceramist. A devout and outspoken **Huguenot**, he was imprisoned for his religious beliefs and for his involvement in the Protestant riots of the first instances of the **Wars of Religion**. It was only with the help of his influential Catholic patron, **Anne de Montmorency**, that he obtained amnesty. Catherine de Medici later acted as his protector, commissioning Palissy to build a private grotto for her at the garden of the Tuileries palace.

Beginning in 1575, Palissy gave public lectures in Paris on natural history that, when published as *Discours Admirables* (*Admirable Discourses*), became extremely popular and revealed him as both a writer and experimental pioneer. In 1588, as the struggle against the Protestants grew, Palissy was again imprisoned. He died two years later of starvation and maltreatment.

QUESTIONS FOR TEACHING

- Name the animals you see on this object.
- What shapes do you see?
- Where do you see repetition?
- Which shapes, colors, or textures are balanced (or distributed evenly) in the composition?
- How are the animals arranged on the object?
- Where might you find all these animals in nature?
- What do the animals have in common?
- This object is known as a basin and might have held water. Imagine this basin filled with water. Can you guess what type of environment it is trying to mimic? Consider the relationships of the plants, animals, and water.
- What would you use this basin for?

LESSON PLAN | BEGINNING LEVEL

Grades: Lower elementary (K–2), upper elementary (3–5)
Subjects: Science and visual arts
Time required: 3–4 class periods

Lesson Overview

Students study how **balance** is created by the arrangement of plants and animals on a sixteenth-century **ceramic** basin. They classify the plants and animals depicted on the basin as **herbivores, carnivores**, or **omnivores**. They research an **ecosystem** to depict on their own visually balanced decorative plates.

Learning Objectives

Students will:

- Identify repetition and balance in a decorative art object.
- Identify the animals in an ecosystem and classify them as herbivores, carnivores, or omnivores based on physical characteristics and prior knowledge.
- Research the plants and animals in a selected ecosystem.
- Create a decorative plate that demonstrates a balance of **shapes**, **colors**, or textures.

Materials

- Image of the oval basin by Bernard Palissy (p. 8)
- Information about the featured work of art and Questions for Teaching (pp. 26–27)
- Copies for students of information about the featured work of art
- Journals or bound paper
- Student handout: Categorizing Life Forms (p. 30)
- Student handout: Herbivores, Carnivores, and Omnivores (p. 31)
- Key to student handouts: Categorizing Life Forms (p. 36)
- Internet access or life science texts for research
- Plastic plates
- Self-hardening clay
- Rolling pin or glass jar
- Forks and popsicle sticks, or other tools to use with clay
- A variety of small plastic animals and plants

Lesson Steps

1. Divide the class into working groups of two to three students. Display an image of Bernard Palissy's oval basin and ask students to share their initial observations. Have the groups make lists in their journals of the life forms they see depicted on the basin. Students should also keep track of how many of each of the life forms are visible on the basin.

2. As a class, review the lists and categorize the life forms into plants and animals using the Categorizing Life Forms student handout.

3. Each group should try to determine what type of environment each animal and plant usually inhabits by using prior knowledge and details in the artwork. Students should brainstorm together and record their hypotheses in their journals. Ask students to think about where they have seen these animals in real life. Each group will share its results with the class. Make a comprehensive list of all the groups' hypotheses and clarify which ideas are correct. Discuss how incorrect answers may have been based on correct observations. For instance, one may guess that this is an ocean environment. This could be a result of seeing the large crawfish, which looks very similar to a lobster.

4. Explain that basins like this would have been used to hold water. Have students describe the various colors and textures used in the background of the basin. Students should notice the patches of brown and the smooth yellow surface. Have the class imagine this basin filled with water. Ask students to describe any ecosystems the basin may resemble when it is filled with water. Ask students to identify which of these ecosystems includes all of these life forms. After the discussion, identify the ecosystem of the basin as a **marshland**.

5. Explain the terms *herbivore, carnivore,* and *omnivore.* Using the Herbivores, Carnivores, and Omnivores student handout, have the groups classify the animals in the basin based on what they see and what they know about each animal's habits and diets within a marshland ecosystem.

6. Ask each group to research the diet of one animal depicted on the basin in this marshland ecosystem and record findings in their journals. Groups will share their results with the class. Students can modify their classification of the animals in step 5 based on class findings.

7. Return to the image of the oval basin and ask students the following questions:
   - What shapes do you see?
   - Where do you see repetition?
   - Which shapes, colors, or textures are balanced (or distributed evenly) in the **composition**?
   - How are the animals arranged on the object?

8. Students will make their own decorative plates, bowls, or basins in the style of Palissy by creating balance through the placement of repeated animals and plants in their compositions. First, students should identify specific ecosystems for their subjects. Then they will research the different types of plants and animals that live in those environments and include them in their final compositions.

9. Give each student a sturdy plastic plate. Have students take pieces of clay, approximately the size of tennis balls, and roll them out into a pancake shape using a rolling pin or glass jar. Instruct them to press the clay into their plates, covering the entire surface. Allow students to decorate the perimeters of their plates with patterns of their choice. They can add pieces of clay (additive process) or scrape or cut into the clay using forks or popsicle sticks (subtractive process). Students then use store-bought plastic plants and animals to decorate their serving pieces. Tell them to experiment with the placement of the plants and animals until they create balance. Have them press the plastic pieces into the clay while the clay is still moist. Once the decorative serving pieces are dry, display them around the room. Students should look closely at each piece and identify which **habitat** each artwork represents based on the life forms depicted and discuss the textures and overall compositions.

STUDENT HANDOUT

CATEGORIZING LIFE FORMS

Use this graphic organizer to classify the objects in Palissy's oval basin as plants or animals. Also note the number of each type of plant or animal in the artwork.

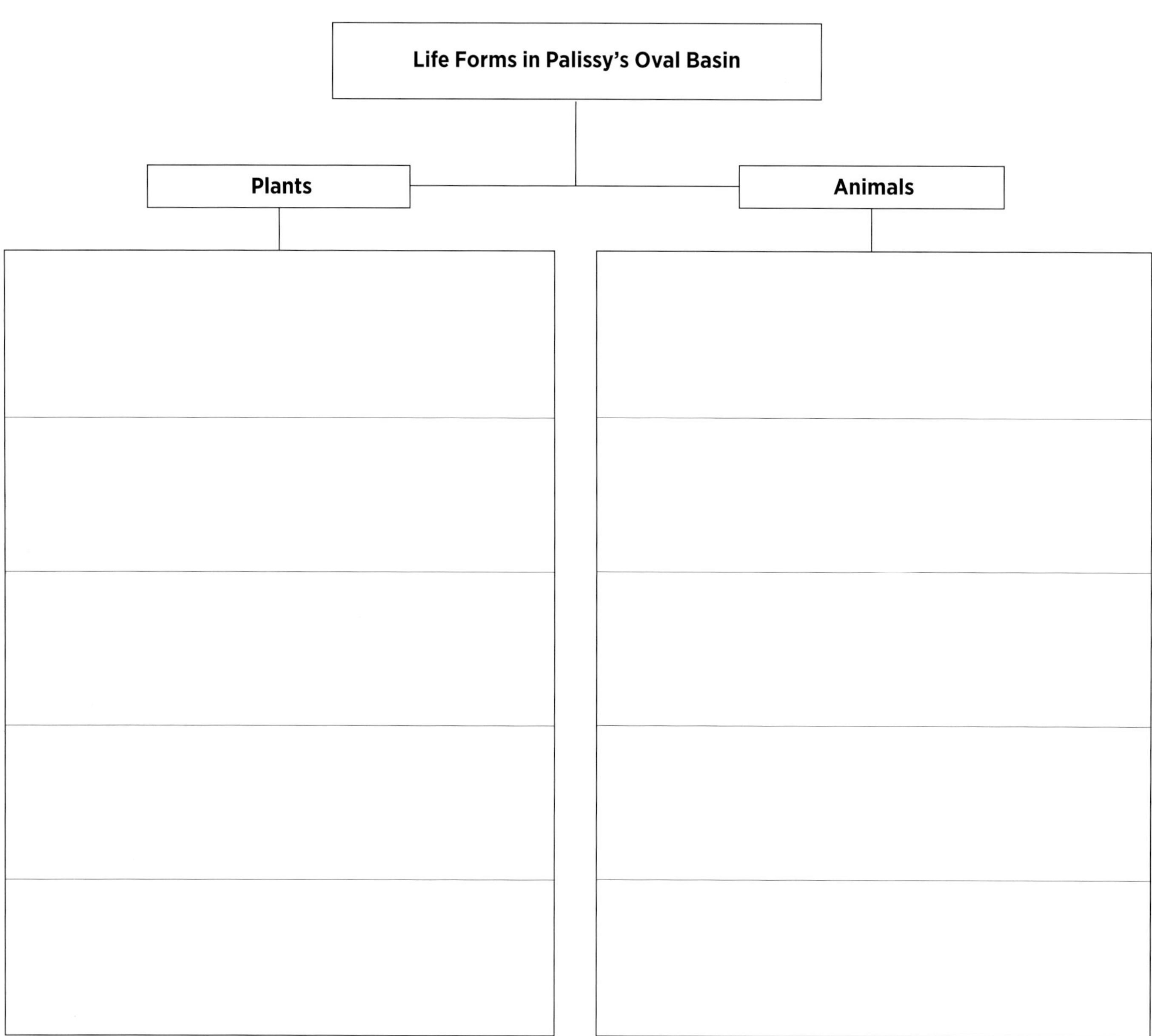

STUDENT HANDOUT

HERBIVORES, CARNIVORES, AND OMNIVORES

Use this graphic organizer to classify the animals in Palissy's oval basin based on what you see and know about each animal's habits and diet in a marshland ecosystem.

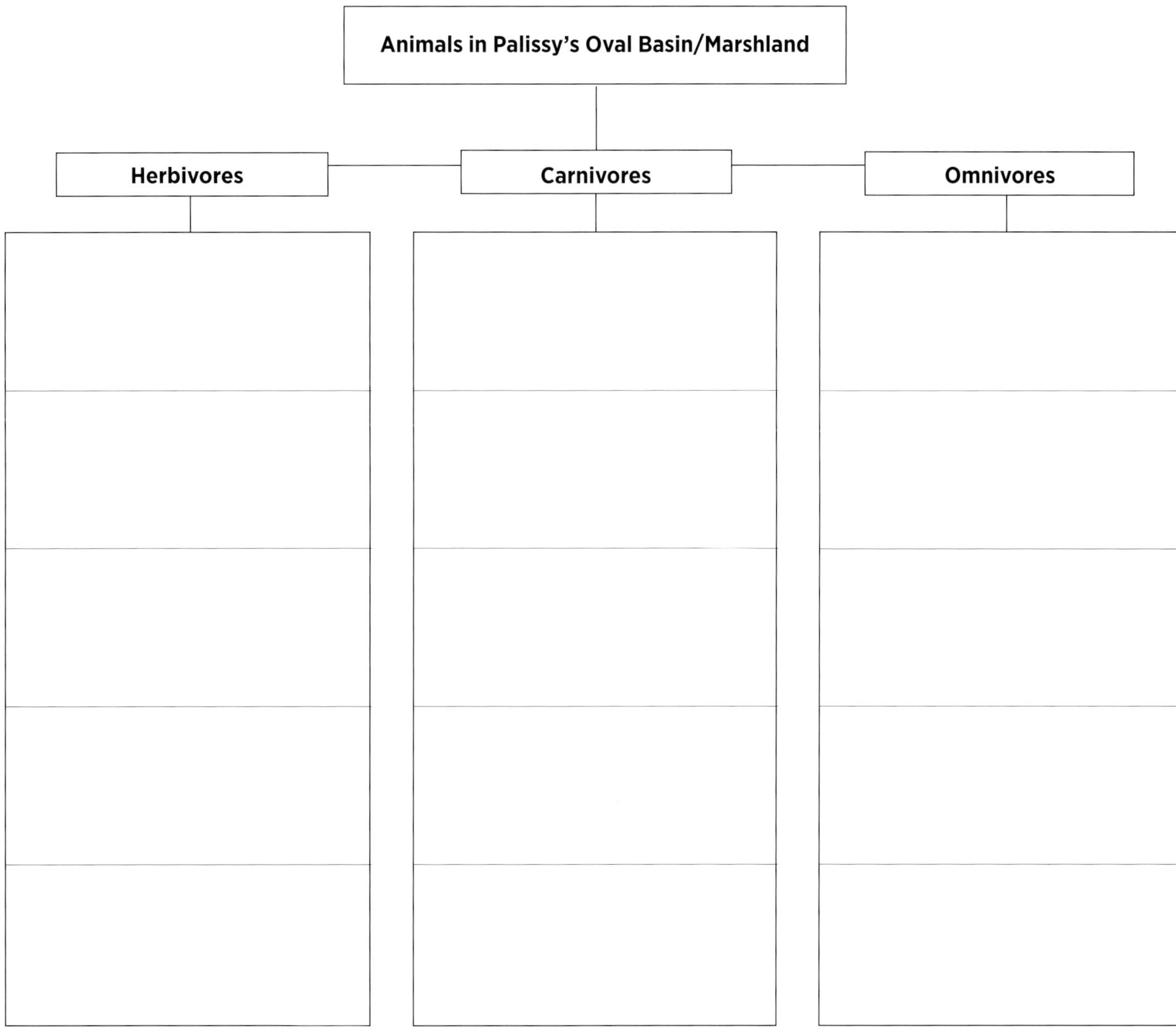

LESSON PLAN | INTERMEDIATE LEVEL

Grades: Middle school (6–8)
Subjects: Science and visual arts
Time required: 4–5 class periods

Lesson Overview

Students study how **balance** is created by the arrangement of plants and animals on a sixteenth-century **ceramic** basin. They classify the plants and animals depicted on the basin based on diet and whether they are **producers**, **consumers**, or **decomposers**. They make their own plates depicting living organisms in a selected **ecosystem** and demonstrate their understanding of visual balance. They create **food webs** based on both the sixteenth-century basin and their own decorative plates.

Learning Objectives

Students will:

- Identify repetition and balance in a decorative art object.
- Research the plants and animals in a selected ecosystem.
- Identify the animals in an ecosystem and classify them as **herbivores**, **carnivores**, or **omnivores** based on physical characteristics, research, and prior knowledge.
- Determine which life forms in an ecosystem are producers, consumers, and decomposers and create a food web that shows transfer of matter within the ecosystem.
- Sculpt a decorative clay plate that demonstrates a balance of **shapes**, **colors**, or textures.

Materials

- Materials listed in the beginning-level lesson (p. 28)
- Student handout: Producers, Consumers, and Decomposers (p. 34)
- Paper and colored pencils
- Instructions on creating a food web. See, for example, the lesson "Casual Patterns in Ecosystems" on Harvard University's Understanding of Consequences Teacher Resource website, www.pz.harvard.edu/ucp/curriculum/ecosystems/s1_res_weboflife_prep.htm

Lesson Steps

1. Complete step 1 and steps 3–6 of the beginning-level lesson.

2. Introduce the terms *producer, consumer,* and *decomposer*. Explain that consumers can function as **primary**, **secondary**, and **tertiary consumers** depending on what they consume.

3. Assign each working group one of the nine life forms depicted in the basin. The members of each group will determine if their life form is a producer, a primary or secondary consumer, or a decomposer, using the definitions and their findings from the beginning-level lesson. They should record their findings on the Producers, Consumers, and Decomposers student handout. Students can use prior knowledge and conduct research to form conclusions about their life forms. Groups will present their findings to the class.

4. Students will apply their knowledge of producers, consumers, and decomposers by creating a food web for the **marshland** ecosystem depicted on the oval basin. Students will draw and label each type of life form in the basin/marshland ecosystem using colored pencils. Each web should include primary, secondary, and tertiary consumers. Each web should also indicate the transfer of matter. Students may sketch webs in their journals before drawing their final web on large drawing paper. See instructions on creating a food web. For example, view the lesson "Casual Patterns in Ecosystems" on Harvard University's Understanding of Consequences Teacher Resource website.

5. Complete steps 7–8 of the beginning-level lesson. Next, have students identify whether the plants and animals in their selected ecosystems are producers, consumers, or decomposers.

6. Have students use self-hardening clay to create plants and animals that live in their selected ecosystems. Encourage them to use both additive processes (e.g., adding pieces of clay) and subtractive processes (e.g., scraping or cutting into the clay) to create a variety of forms and textures. You may wish to reference the following lesson plans on the Getty website that use clay to create sculptures:
   - "Python, Python, What Do You See?" www.getty.edu/education/teachers/classroom_resources/curricula/arts_lang_arts/a_la_lesson31.html
   - "Capturing the Moment in 3-D" www.getty.edu/education/teachers/classroom_resources/curricula/sculpture/lesson01.html

7. Give each student a sturdy plastic plate. Have students take pieces of clay, approximately the size of tennis balls, and roll them out into a pancake shape using a rolling pin or glass jar. Instruct them to press the clay into their plates, covering the entire surface. Have students arrange their plant and animal sculptures on their plates, experimenting with the placement of the sculptures until they are balanced. Have them score the bottom of each sculpture and the part of the plate where they will place each plant or animal. Have them gently press the clay pieces onto the plates while the clay is still moist.

8. When their decorative plates are dry, students can use **glaze** or acrylic paint to add color and texture to their plates to create unified **compositions**. Once the plates are completed, students should create a food web based on the plants and animals depicted in their own decorative art objects.

9. Display the plates around the room. Students should look closely at each piece and identify which **habitat** each artwork represents, based on the life forms depicted, and discuss the colors, textures, and overall composition.

STUDENT HANDOUT

PRODUCERS, CONSUMERS, AND DECOMPOSERS

Define the terms *producer, consumer,* and *decomposer.*

Consumers can function as primary, secondary, and tertiary consumers depending on what they consume. Use the definitions of these terms and what you know about the feeding habits of each creature to categorize the life forms in Palissy's oval basin into the four areas below.

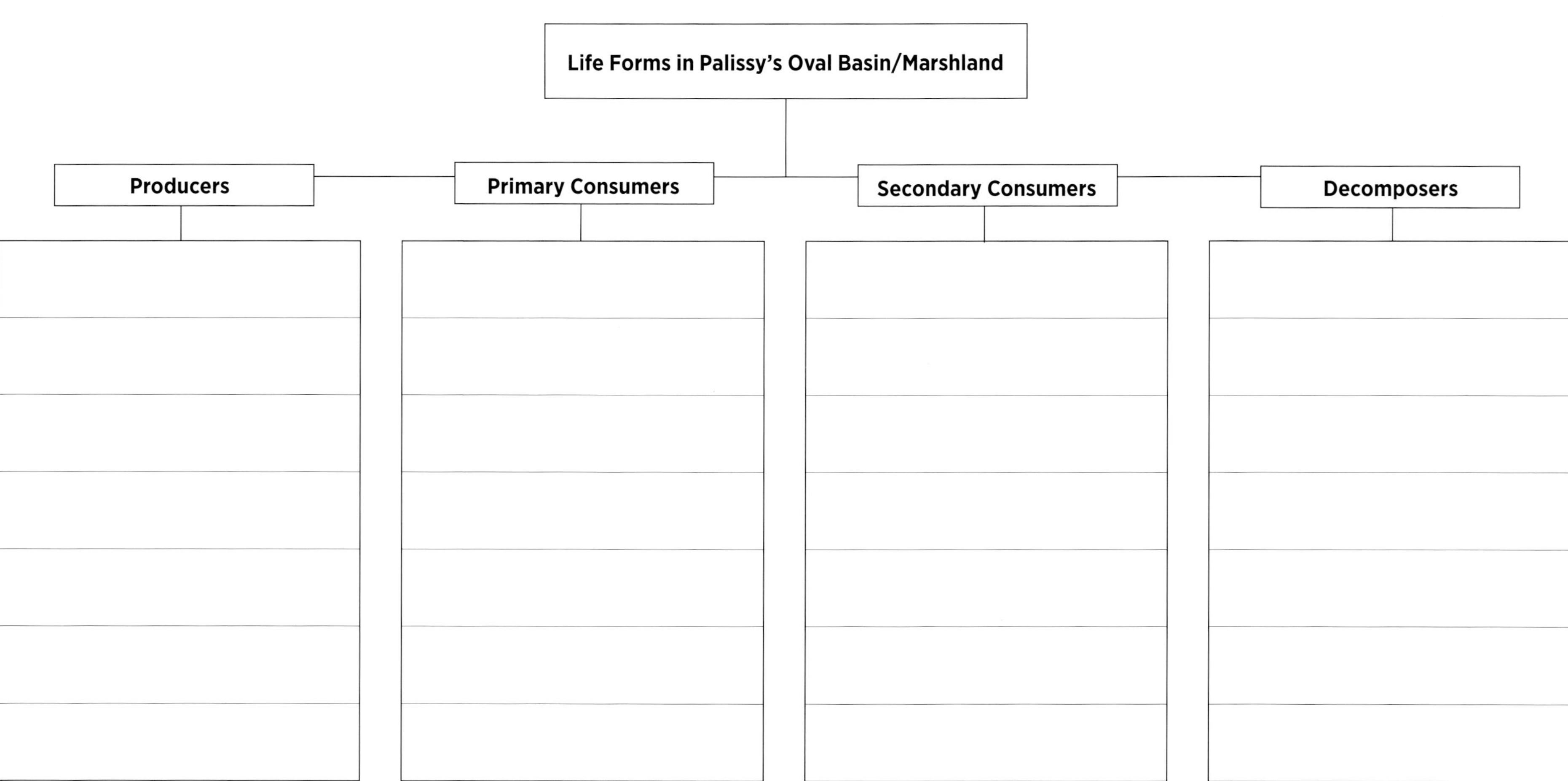

LESSON PLAN | ADVANCED LEVEL

Grades: High school (9–12)
Subjects: Science and visual arts
Time required: 4–5 class periods

Lesson Overview

Students study how **balance** is created by the arrangement of plants and animals on a sixteenth-century **ceramic** basin. They classify the plants and animals depicted on the basin based on diet and whether they are **producers**, **consumers**, or **decomposers**, and they identify dominant species in the **marshland ecosystem**. They make their own plates depicting living organisms in a balanced ecosystem and demonstrate their understanding of visual balance. They create **food webs** based on both the sixteenth-century basin and their own decorative plates.

Learning Objectives

Students will:

- Identify the animals in an ecosystem and classify them as **herbivores**, **carnivores**, or **omnivores** based on physical characteristics, research, and prior knowledge.
- Determine which life forms in an ecosystem are producers, consumers, and decomposers and create a food web that shows transfer of matter within the ecosystem.
- Analyze a food web, predict which life forms will influence an ecosystem, and predict how that ecosystem might change due to the influence of additional species.
- Sculpt a decorative clay plate that depicts a balanced ecosystem and demonstrates a balance of **shapes**, **colors**, or textures.

Materials

- Materials listed in the beginning-level lesson (p. 28)
- Student handout: Producers, Consumers, and Decomposers (p. 34)
- Colored pencils

Lesson Steps

1. Display an image of Bernard Palissy's oval basin. Count the number of each type of life form depicted in the basin. Review each life form's role in the ecosystem by completing steps 3–6 of the beginning-level lesson and steps 2–4 of the intermediate-level lesson.

2. Discuss the term *biodiversity*. Explain that biodiversity refers to the number and variety of species in an environment. Biodiversity is affected by both **biotic** and **abiotic** systems. The biotic systems of the marshland have just as much effect on the **habitat** as the abiotic systems in the water, mud, and air in the marshland.

3. Explain that the habitat depicted in the basin is an unbalanced system. Discuss which animals or plants are dominant. Students should refer to the animals' placement in the food web and consider how many are depicted. As a class, brainstorm about what would happen if the basin were an actual environment. Which species could have the most effect on the other populations? In other words, if this were a real habitat, who would survive? Students should record predictions in their journals.

4. Analyze the food webs students created and discuss which life forms would dominate these ecosystems. Students will predict how the ecosystems might be altered by changes in both biotic and abiotic systems. Determine if changes would be desirable. What would need to be added to or subtracted from the habitat to keep it balanced? Ideas may range from changes in the number of a given species to introducing mates or new food sources into the environment. Using colored pencils, students will use the results of the discussion to create a sketch in their journals for a basin that has a balanced environment. Students should pay attention to the placement of animals as well as relative size to create visual balance. Students will share their ideas with the class.

5. Complete steps 7–8 of the beginning-level lesson. Students should include abiotic factors in their **compositions**. Students will draw two or three designs for their own decorative plates in their journals.

6. Complete steps 6–9 of the intermediate-level lesson, adapting the steps to focus on depicting a balanced ecosystem on the decorative plates.

KEY TO STUDENT HANDOUTS

CATEGORIZING LIFE FORMS

Plants	Animals
Fern (2)	Clam (12)
Seaweed—spiked leaf (3)	Fish (2)
Seaweed—round leaf (16)	Lizard (3)
	Snake (2)
	Crawfish (2)
	Turban Snail (24)
	Frog (3)
	Tower Screw Snail (12)

HERBIVORES, CARNIVORES, AND OMNIVORES

Herbivores	Carnivores	Omnivores
Fish	Snakes	Frogs
Snails	Lizards	Crawfish
Clams		

PRODUCERS, CONSUMERS, AND DECOMPOSERS

Producers	Primary Consumers	Secondary Consumers	Decomposers
Ferns	Fish	Snakes	Snails
Seaweeds	Frogs	Lizards	Clams
		Crawfish	

Drawn to Biology
Nature Classified

Students learn that observation skills are important in the fields of art and science. They closely observe plants and insects in nature and in art and make drawings based on their observations. Students classify the plants and insects depicted in a seventeenth-century still-life painting.

Beginning Level
Students discuss and describe a still-life painting and categorize the things they see in the painting as living or nonliving. They identify lines and shapes in depictions of insects and observe and draw details from the painting.

Intermediate Level
Students discuss the lines, shapes, and other formal elements of art observed in a seventeenth-century still-life painting depicting flowers and insects. They use lines, shapes, and value to create observational drawings of flower or insect specimens they collect. They learn about Linnaeus's system of scientific names and then research the scientific names of their specimens.

Advanced Level
Students discuss the flowers and insects depicted in a seventeenth-century still-life painting. They use lines, shapes, and value to create realistic depictions and employ these strategies in their own drawings of specimens and diagrams of cells. They learn about Linnaeus's system of scientific names, research the scientific names of their specimens, and discover how new species can arise.

Standards Addressed
Refer to the charts for national and California state standards on the Getty website, www.getty.edu/education/teachers/classroom_resources/curricula/art_science/downloads/standards.pdf

FEATURED WORK OF ART ►

Ambrosius Bosschaert the Elder
Dutch, 1573–1621
Flower Still Life, 1614
Oil on copper
11¼ x 15 in.
J. Paul Getty Museum
83.PC.386
www.getty.edu/art/gettyguide/artObjectDetails?artobj=842

In this painting, a pink carnation, a white rose, and a yellow tulip with red stripes lie in front of a basket of brilliantly colored blossoms. Various types of flowers that would not bloom in the same season appear together here: rose, forget-me-not, lily-of-the-valley, cyclamen, violet, hyacinth, and tulip. Rendering meticulous detail, Ambrosius Bosschaert the Elder conveyed the silky texture of the petals, the prickliness of the rose thorns, and the fragility of opening buds. **Insects** crawl, alight, or perch on the bouquet. Each is carefully described and observed, from the dragonfly's transparent wings to the butterfly's minutely painted **antennae**. Although a vague reference, insects and short-lived flowers are a reminder of the brevity of life and the transience of its beauty.

A rising interest in **botany** and a passion for flowers led to an increase in painted floral still lifes at the end of the 1500s in both the Netherlands and Germany. Bosschaert was the first great Dutch specialist in fruit and flower painting and the head of a family of artists. He established a tradition that influenced an entire generation of fruit and flower painters in the Netherlands.

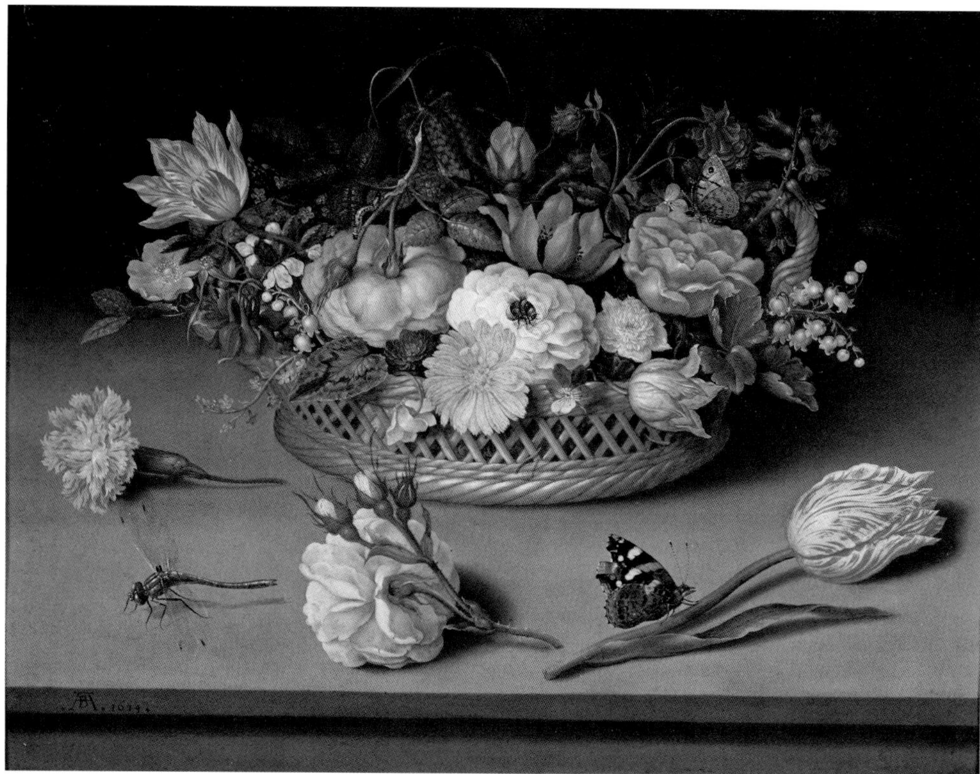

ABOUT THE ARTIST

During the 1600s the Dutch became Europe's leading **horticulturists**, and exotic flowers became a national obsession. Not surprisingly, flower painters were among the best-paid artists. In 1621, Ambrosius Bosschaert the Elder commanded a thousand **guilders** for a single flower picture. Nonetheless, his output of artworks was relatively small, for he was by trade an art dealer. Anticipating religious persecution, in 1587 Bosschaert's parents moved from Antwerp to Middelburg, a seaport and trading center second in importance only to Amsterdam. Six years later, Bosschaert joined Middelburg's **Guild of Saint Luke**. Bosschaert's works have been called flower portraits; each flower receives the same detailed attention as a face in a portrait. Usually small in scale and on **copper**, Bosschaert's paintings combined blooms from different seasons, painted from separate studies of each flower. It is not unusual to find the same flower, shell, or insect in many of his pictures. Like his predecessors, Bosschaert sometimes included symbolic or religious meanings in his works, such as the transience of life, by including objects at different stages in the life cycle.

QUESTIONS FOR TEACHING

- What objects do you see that are from nature? What objects do you see that are man-made?
- How many different kinds of flowers do you see?
- What are some reasons that an artist might have chosen to paint this subject?
- Artists in Bosschaert's time were very interested in all aspects of the natural world. What evidence of this can you find in this painting?
- How does the inclusion of insects in the composition affect your interpretation of the painting?

LESSON PLAN | **BEGINNING LEVEL**

Grades: Lower elementary (K–2), upper elementary (3–5)
Subjects: Science and visual arts
Time required: 2 class periods

Lesson Overview

Students discuss and describe a still-life painting and categorize the things they see as living or non-living. They identify **lines** and **shapes** in depictions of **insects** and observe and draw details from the painting.

Learning Objectives

Students will:

- Categorize living and non-living subjects in a still-life painting.
- Identify lines and shapes in a painting.
- Chart classifying characteristics of insects.
- Closely observe insects in an artwork and draw them using a variety of lines and shapes.

Materials

- Reproduction of *Flower Still Life* by Ambrosius Bosschaert the Elder (p. 9)
- Information about the featured work of art and Questions for Teaching (p. 37–38)
- Five detail reproductions of insects from the painting: www.getty.edu/education/teachers/classroom_resources/curricula/art_science/00084201_butterfly1.html www.getty.edu/education/teachers/classroom_resources/curricula/art_science/00084201_butterfly2.html www.getty.edu/education/teachers/classroom_resources/curricula/art_science/00084201_bumblebee.html www.getty.edu/education/teachers/classroom_resources/curricula/art_science/00084201_dragonfly.html www.getty.edu/education/teachers/classroom_resources/curricula/art_science/00084201_caterpillar.html
- Art activity: "Drawing Insects with Organic Shapes and Lines," www.getty.edu/education/teachers/classroom_resources/tips_tools/downloads/drawing_insects.pdf
- Prepared notes about the **lifespan**, **habitat**, and biology of the insects visible in the painting: butterfly, bumblebee, dragonfly, and caterpillar
- Student handout: Insect Info (p. 40)

Lesson Steps

1. Display or hand out reproductions of *Flower Still Life* by Ambrosius Bosschaert the Elder. Ask students the following questions:
   - What do you see? What more can you find?
   - What kinds of lines do you see?
   - What kinds of shapes do you see?

2. Tell students that they are looking at a still-life painting. Still-life paintings often depict living and non-living things together—animals, plants, and objects. Ask students to identify things that are alive, and then have them identify non-living objects. Create a graphic organizer on the board to classify objects in the artwork as either living or non-living. Then, within the list of living things, have the class identify the plants and animals in the still life and sort them into the graphic organizer. Tell students that classification is an important step in understanding all the things that make up our world. Scientists classify things by finding common characteristics. By sorting the items in this still life by whether they are living or non-living, and then by whether they are plants or animals, the class has created a classification of the items in the painting.

3. Give each pair of students a reproduction of a detail of a different insect from the painting. Remind students of the lines and shapes they noticed in the depictions of the insects in the painting (e.g., ovals for the head and body, thin lines for the legs). Have students draw the insects using a variety of lines and shapes. For illustrated, step-by-step drawing instructions, view the art activity "Drawing Insects with Organic Shapes and Lines" on the Getty website.

4. Have partners answer the questions about their insect on the Insect Info student handout, using their observations and the notes you prepared for the students.

5. Once the students have answered the questions, chart the answers on the board. Ask how this list could be divided into two subgroups of insects that share certain qualities. Some options include: whether they live in groups or by themselves, winged and non-winged insects, and **carnivores** and **herbivores.** Come up with as many variations as possible.

STUDENT HANDOUT

INSECT INFO

With a partner, answer the questions below about your picture of an insect.
Your answers should be based on information your teacher gives you about the insect.
You may also use information about the artwork to help you.

1. How long does the insect usually live?

2. What does it eat?

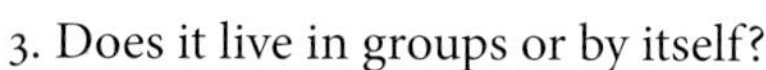

3. Does it live in groups or by itself?

4. What kind of environment does it live in?

5. In what ways does the insect move from place to place?

LESSON PLAN | **INTERMEDIATE LEVEL**

Grades: Middle school (6–8)
Subjects: Science and visual arts
Time required: 2 class periods

Lesson Overview

Students discuss formal elements used in a seventeenth-century still-life painting depicting flowers and **insects**. They create observational drawings of flower or insect **specimens** they collect. They learn about **Linnaeus**'s system of scientific names and research the scientific names of their specimens.

Learning Objectives

Students will:

- Understand Linnaeus's system of classification.
- Draw insects and flowers from close observation using techniques observed in a still-life painting.
- Understand that observation skills are important in the fields of art and science.

Materials

- Materials listed in the beginning-level lesson (p. 39)
- Bug viewers or magnifying glasses
- Paper, pencils, and colored pencils
- Art activity: "Exploring Value to Create Form," www.getty.edu/education/teachers/classroom_resources/tips_tools/downloads/exploring_value.pdf
- Reference resources for Linnaeus's system of scientific names. Some online resources for ideas about how to use keys to help classify plants and animals include:
  BugGuide, bugguide.net
  Royal Botanic Gardens, Kew, www.kew.org/education/index.html
  Flora of North America lesson, "Species and Specimens," floranorthamerica.org/Outreach/FNA_lesson_biodiversity

Lesson Steps

1. Complete step 1 of the beginning-level lesson, adapting for grade level as appropriate.

2. Explain to the class that an interest in **botany** and a passion for flowers led to an increase in painted floral still lifes at the end of the 1500s in both the Netherlands and Germany. Paintings by Ambrosius Bosschaert the Elder were called flower portraits because each flower received the same detailed attention as a face in a portrait. Ask students what the artist did to make the flowers and insects look three-dimensional. Point out examples of shadows and light reflection in the painting. Point out the importance of observational drawing in both art and science. Tell students that Bosschaert carefully observed flowers and insects in order to draw them with meticulous detail.

3. Have each student collect an insect or flower specimen from an area of the schoolyard or from home and bring it to class. Have students write down what they notice about the specimen.

4. Review with students the reasons for classification and background on Linnaeus's system of naming, called **binomial nomenclature**, which was instituted in the eighteenth century. Chart the scientific names of various familiar organisms like *Canis familiaris* (domesticated dog), *Eschscholzia californica* (California poppy), or *Ursus horribilis* (grizzly bear) and explain the hierarchy implicit in the names, which describe the scientific family and **species**. For example, explain to students that the Latin word *Ursus* means "bear" and is the family name that all bears belong to. The second word, *horribilis,* is Latin for "horrible" and describes the specific type of bear within the family of bears.

5. Hand out bug viewers or magnifying glasses to the students. Students will observe their insects or flowers and draw them carefully. Remind students of the **lines** and **shapes** that are visible in the insects and flowers in the painting (e.g., ovals for the head and body, thin lines for the legs; ovals for petals, long parallel lines for stems). You may wish to view the art activity "Drawing Insects with Organic Shapes and Lines" on the Getty website. Students should use variations in **value** to give the specimens three-dimensional form. You may wish to view the art activity "Exploring Value to Create Form" on the Getty website. Ask students if drawing their specimens helped them to observe a detail they did not notice previously.

6. Using scientific texts, students should research the scientific names of their specimens. The lesson "Species and Specimens" from the Flora of North America website provides ideas about how to use keys to identify plants and animals.

7. Display the correctly labeled drawings around the classroom.

LESSON PLAN | **ADVANCED LEVEL**

Grades: High school (9–12)
Subjects: Science and visual arts
Time required: 2–3 class periods

Lesson Overview

Students discuss the flowers and **insects** in a seventeenth-century still-life painting. They identify artistic techniques used to create realistic depictions and employ these techniques in their own drawings of **specimens** and diagrams of **cells**. They learn about **Linnaeus**'s system of scientific names, research the scientific names of specimens, and discover how some new plant varieties can arise.

Learning Objectives

Students will:

- Understand Linnaeus's system of classification.
- Understand that new plant varieties can arise due to viruses.
- Understand that observation skills are important in the fields of art and science.
- Draw diagrams of plant cells and viruses, using techniques observed in a still-life painting.
- List at least three characteristics or properties of a virus.

Materials

- Materials listed in the beginning-level lesson (p. 39)
- Art activity: "Exploring Value to Create Form," www.getty.edu/education/teachers/classroom_resources/tips_tools/downloads/exploring_value.pdf
- Research materials on viruses
- Botanical texts and other research tools, such as:
  BugGuide, bugguide.net
  Royal Botanic Gardens, Kew, www.kew.org/education/index.html
  Flora of North America lesson, "Species and Specimens," floranorthamerica.org/Outreach/FNA_lesson_biodiversity
- Pencils and colored pencils

Lesson Steps

1. Display or hand out reproductions of *Flower Still Life* by Ambrosius Bosschaert the Elder. Have students share their initial observations about the painting, then ask them the following questions:
   - What can you identify in this painting?
   - What are some reasons that an artist might have chosen to paint this subject?
   - Artists in Bosschaert's time were very interested in all aspects of the natural world. What evidence of this can you find in this painting?

2. Complete steps 2–4 and 6 of the intermediate-level lesson (p. 41).

3. Explain to students how new classifications and variations of living things can arise. For information on evolution and genetic mutation visit the BBC GCSE Bitesize website, www.bbc.co.uk/schools/gcsebitesize/science/21c/life_on_earth/theory-evolutionrev1.shtml.

4. Return to the image of the painting. Have students focus on the tulips in the painting. Ask them to identify differences between the tulips and the other flowers depicted.

5. Explain to students that viruses caused the stripes and bright color streaks in the tulips seen in the painting. These viruses once made tulips valuable commodities. In seventeenth-century Holland, infected tulips were highly prized, and people traded enormous fortunes for the thrill of owning even one infected bulb.

6. Review with your students what they know about viruses. Point out the fact that they occur in plants as well as animals.

7. Have students use printed research materials and the Internet to add to their knowledge about viruses in tulips. Students should understand the following before they continue with the activity:
   - A virus is an infectious organism that reproduces within the cells of a host.
   - A virus is not alive until it enters the cells of a living plant or animal.
   - Viruses can be introduced in plants by insects, other plant seeds, or pollen.
   - A virus is composed of genetic information within a protein coat.
   - Viruses can be contained so that they do not infect the environment or other organisms.

8. Ask students to draw a diagram of a plant cell and a virus, based on the research they just completed. Students should use variations in **value** to distinguish the different components of the cell and the virus in their drawings. For further instructions, you may wish to view the art activity "Exploring Value to Create Form" on the Getty website. Write out a list of similarities and differences comparing the plant cell and the virus.

9. Divide the class into groups. Tell group members that they are tulip farmers who want to create a new variety of tulip by genetically engineering a virus and introducing it to a crop. Ask students to address the following based on the research they have done in previous steps:
   - Describe how they will introduce the virus.
   - Predict how the infected tulips will look.
   - Determine how they will safeguard against the virus spreading into the environment and contaminating other plants.

Stars in Space, Mythology, and Manuscripts

Students explore constellations by viewing and discussing a page from an astronomy textbook made hundreds of years ago. They chart constellations, identify the characteristics of the stars that compose them, and contribute to a class astronomy book.

Beginning Level
Students learn the definition of *constellation*. They focus on the Ursa Major constellation, read the associated myth, and then draw invented constellations based on random clusters of paper stars. Students write their own myths based on their constellations and include their myths and drawings in a class book.

Intermediate Level
Students learn about the Ursa Major constellation and then conduct independent research on constellations of their choice. They combine text and images on pages that depict and describe their constellations and contribute their pages to a class astronomy textbook.

Advanced Level
Students conduct independent research on constellations of their choice, focusing on temperature and luminosity. Students manipulate text and images in a graphics software program to contribute pages to a class digital astronomy textbook.

Standards Addressed
Refer to the charts for national and California state standards on the Getty website, www.getty.edu/education/teachers/classroom_resources/curricula/art_science/downloads/pdf

FEATURED WORK OF ART ►

Astrological Cycle with Commentary
From the manuscript *Astronomical Miscellany*
English, early 1200s
Black, green, and red inks on parchment
9½ x 6⅛ in.
J. Paul Getty Museum
Ms. Ludwig XII 5, fol. 149v
www.getty.edu/art/gettyguide/artObjectDetails?artobj=136048

The basic course of learning in the Middle Ages was the study of the seven liberal arts: grammar, rhetoric, logic, arithmetic, music, geometry, and **astronomy**. A renewed interest in the natural world in the 1200s ensured a prominent place for astronomy in the growing universities of Europe. This page comes from a collection of scientific theories about the **constellations** and other scholarly texts that were probably compiled as a textbook. The constellations and signs of the zodiac were rendered in a pen-and-ink technique that lends liveliness to the depiction of the positions of the stars in the sky.

On this page can be seen a number of constellations, such as **Ursa Major** (the great bear), Ursa Minor (the little bear), Hercules (holding a club and lionskin), and Draco (the dragon).

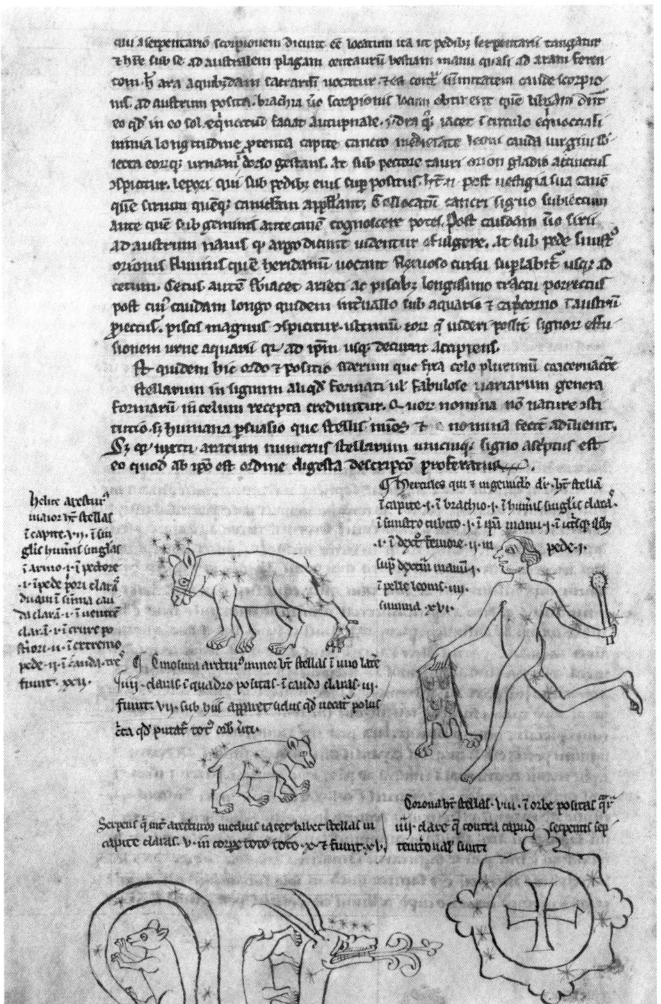

THE STORY OF URSA MAJOR

Calypso was a nymph and a friend of Artemis, the Greek goddess of the hunt. Calypso and Artemis spent lots of time together in the forest hunting and having adventures. When Zeus, the king of all the gods, happened to see Calypso, he fell in love with her. He appeared before her and told her that he loved her, and she fell in love with him as well. The trouble was that Hera, Zeus's wife and the queen of all the gods, had a very bad temper. Zeus disguised Calypso as a bear so that the jealous Hera would not find her.

One day when Artemis was out hunting, she saw a beautiful she-bear and shot it with her bow and arrow. She did not realize that it was her friend Calypso! Zeus arrived in the nick of time and took the wounded bear up to the heavens to keep her safe. According to this myth, when you look up at the northern sky at night, you can see the bright stars of Ursa Major (the great bear), once the nymph Calypso, twinkling down.

QUESTIONS FOR TEACHING

- What do you see on this page?
- Do any of the images on this page look familiar?
- What other kinds of books with illustrations and short descriptions have you used before? What was the purpose of the book?
- What were possible purposes for these illustrations?
- How do constellations play a role in society today?
- Astronomy and astrology were considered to be the same thing at the time this book was made. How do we view them today?

LESSON PLAN | **BEGINNING LEVEL**

Grades: Lower elementary (K–2), upper elementary (3–5)
Subjects: Science, visual arts, and English language arts
Time required: 2–3 class periods

Lesson Overview

Students learn the definition of *constellation* by viewing images in an astronomical textbook made hundreds of years ago. They learn about the **Ursa Major** constellation and associated myth and then draw invented constellations based on random clusters of paper stars. Students write their own myths based on their constellations and include their myths and drawings in a class book.

Learning Objectives

Students will:

- Learn the definition of *constellation*.
- Speculate on the reason humans have organized star clusters into constellations.
- Draw invented constellations in the shapes of animals or other creatures.
- Write original myths associated with their constellations.

Materials

- Reproduction of a page from the **manuscript** *Astronomical Miscellany* (p. 10)
- Information about the featured work of art and Questions for Teaching (pp. 44–45)
- Student handout: Ursa Major's Stars (p. 48)
- Image of Ursa Major (e.g., diagram from the Montréal Planetarium website, page 10 on www.planetarium.montreal.qc.ca/Education/Fiches/PDF/constellation_an.pdf)
- Black construction paper
- White or silver paper stars
- Glue sticks
- White chalk or pastels
- (Optional) Art activity: "Introducing Shape," www.getty.edu/education/teachers/building_lessons/introducing_shape.pdf
- Paper and pencils
- (Optional) Video: *Making Manuscripts*, www.getty.edu/art/gettyguide/videoDetails?segid=372
- (Optional) Google Earth software, www.earth.google.com
- (Optional) Art activity: "A Simple Sketchbook," www.getty.edu/education/teachers/classroom_resources/tips_tools/downloads/aa_simple_sketchbook.pdf

Lesson Steps

1. Display the page from the manuscript *Astronomical Miscellany* depicting constellation diagrams, including Ursa Major and Ursa Minor. Ask students what they see. Continue the discussion with Questions for Teaching as appropriate to grade level. Chart responses.

2. Through discussion, ask/tell students what a constellation is. Ask students to speculate about why people would need constellations. Explain that constellations help astronomers to organize the stars in the sky and to identify the part of the sky they are viewing.

3. Ask students to speculate about the type of book this page comes from. Explain that it comes from an astronomical textbook that was written about eight hundred years ago. This textbook included some of the forty-eight constellations known to the writer at the time. Tell students that today there are eighty-eight constellations recognized by astronomers.

4. Show students the image of the stars that make up Ursa Major on the handout Ursa Major's Stars. Ask them if they see a pattern that would help them remember this group of stars among all the stars in the sky. After discussion, show students an image of the constellation outlined. (One example is on the Montréal Planetarium website.) You may also wish to show students the constellation in a virtual reproduction of our solar system through a free computer program called Google Earth, which you can download to your classroom's computers. In Google Earth you can view galaxies, constellations, and planets by choosing "View" from the main menu and clicking "Switch to Sky."

5. Tell students that the constellations were associated with stories or myths. Share the story of Ursa Major (p. 45).

6. Inform students that they are going to create their own constellations. Give each student a piece of black construction paper and six paper stars. Have students hold the stars in their hands about a foot above the paper, then release the stars so they fall randomly on the page. Students should use glue sticks to adhere the stars to the page where they landed.

7. Using white chalk or pastels, students will draw animals or creatures that incorporate their stars to form invented constellations. You may wish to prepare students to draw animals by having them complete the art activity "Introducing Shape" on the Getty website.

8. Have students write myths of grade-appropriate length that feature their constellations. After completing the first drafts, students should exchange myths with their partners. Partners will suggest descriptive details that can be added to the drafts. Students will then complete their own final drafts.

9. Inform students that *Astronomical Miscellany* was created at a time when every aspect of a book was made by hand. You may wish to have students view the video *Making Manuscripts* on the Getty website. Tell students that their drawings and myths will also become part of a handmade book.

10. Bind the constellations and myths together into a class book. For instructions on a simple binding technique, see the art activity "A Simple Sketchbook" on the Getty website.

STUDENT HANDOUT

URSA MAJOR'S STARS

LESSON PLAN | **INTERMEDIATE LEVEL**

Grades: Middle school (6–8)
Subjects: Science and visual arts
Time required: 1–3 class periods

Lesson Overview
Students view and discuss a page from an **astronomy** textbook made hundreds of years ago. They learn about the **Ursa Major constellation** and then conduct independent research on constellations of their choice. Students successfully incorporate text and images on pages that depict and describe their constellations, and they contribute their pages to a class astronomy textbook.

Learning Objectives
Students will:
- Speculate on the reasons humans have organized star clusters into constellations.
- Research, present background on, and define attributes of existing constellations.
- Create pages that successfully combine text and images for a class book.

Materials
- Materials listed in the beginning-level lesson (p. 46)
- Student handout: Constellation List (p. 51)
- Student handout: Constellation Research (pp. 52–53)
- (Optional) Video: *Making Manuscripts*, www.getty.edu /art/gettyguide/videoDetails?segid=372
- Internet access or science texts for research
- White copy paper
- White colored pencils

Lesson Steps
1. Have students make a "Know—Want to Know—Learned" (KWL) chart about constellations and fill in the first two columns. In the "K" column, students should write what they know about constellations. In the "W" column, students should write what they want to know about constellations.

2. Complete steps 1–5 of the beginning-level lesson (p. 46).

3. Pass out copies of the two student handouts (pp. 51, 52–53). Students will select constellations to research from the Constellation List handout and will use the Constellation Research handout as a guide for research and a place to record their findings.

4. Inform students that *Astronomical Miscellany* was created at a time when every part of a book was made by hand. You may wish to have students view the video *Making Manuscripts* on the Getty website. Inform students that they will be creating a class astronomy textbook that includes each student's research.

5. Return to the page from *Astronomical Miscellany* (p. 10). Ask students to share what they notice about the placement of text and images. Point out that the figure of the large bear is surrounded by a square of negative space, while the human figure is tightly surrounded by text. Ask students to share which treatment they prefer and why.

6. Distribute white copy paper and white or silver paper stars. Tell students they will be experimenting with different layouts for their pages of the astronomy textbook. Have students place the paper stars in similar arrangements to those of the stars in their constellations. Tell them to lightly sketch the animals, creatures, or objects they associate with their constellations. Have students write sentences based on their research about their constellations' histories and unique characteristics, with each sentence written on a different piece of white paper. Have students play with the placement of text and images until they can fit all their text around their constellations in a visually pleasing way.

7. Once students are satisfied with their layouts, distribute black construction paper, glue, white chalk or pastels, and white colored pencils. Tell students to glue their stars in place. Have students use the white chalk or pastels to draw the animals, creatures, or objects they associate with their constellations. Have them use the white colored pencils to write in the text.

8. Bind students' pages together to make a class book. For instructions on a simple binding technique, see the art activity "A Simple Sketchbook" on the Getty website.

9. Have students share their findings. Now that they have conducted their research, they should be able to define *magnitude*, *spectral class*, and *color* in relation to a star's characteristics. (For additional information about these terms, see the advanced projects for students on the Sloan Digital Sky Survey website, cas.sdss.org/dr6/en/proj/advanced/.)

10. Revisit the KWL chart. Tell students to complete the "L" column of the chart by writing down what they learned about constellations.

STUDENT HANDOUT
CONSTELLATION LIST

Constellation	English Name	Constellation	English Name
Andromeda	Andromeda	Gemini	Twins
Aquarius	Water Carrier	Hercules	Hercules
Aries	Ram	Hydra	Water Snake
Cancer	Crab	Leo	Lion
Canis Major	Great Dog	Leo Minor	Little Lion
Canis Minor	Little Dog	Libra	Scales
Capricornus	Goat	Lyra	Lyre
Cassiopeia	Cassiopeia	Orion	Orion
Centaurus	Centaur	Pegasus	Winged Horse
Cepheus	Cepheus	Perseus	Perseus
Cetus	Whale	Pisces	Double Fish
Corvus	Crow	Sagittarius	Archer
Crux	Southern Cross	Scorpius	Scorpion
Cygnus	Swan	Taurus	Bull
Draco	Dragon	Ursa Major	Great Bear
Equuleus	Little Horse	Ursa Minor	Little Bear
Eridanus	River	Virgo	Virgin

STUDENT HANDOUT

CONSTELLATION RESEARCH, PAGE 1

Select a constellation from the Constellation List student handout. Use this two-page worksheet to research your constellation and record your findings. You may use books and the Internet.

Constellation Name: ______________________________

Hemisphere: ______________________________

History:	**Diagram:**

What is special about my constellation?

1.

2.

3.

STUDENT HANDOUT

CONSTELLATION RESEARCH, PAGE 2

Stars in the Constellation

Star Name	Spectral Class	Color/Temp.	Characteristics	Magnitude	Distance from Earth

Sources for my research:

1.

2.

3.

LESSON PLAN | **ADVANCED LEVEL**

Grades: High school (9–12)
Subjects: Science and visual arts
Time required: 3–5 class periods

Lesson Overview

Students discuss a page from an **astronomy** textbook made hundreds of years ago. They conduct research on **constellations**, focusing on temperature and luminosity. Students manipulate text and images in a graphics software program to create pages for a class digital astronomy textbook.

Learning Objectives

Students will:

- Speculate on the reason humans have organized star clusters into constellations.
- Research, present background on, and define attributes of an existing constellation.
- Research stars that compose a constellation and chart their temperature and luminosity on a **Hertzsprung-Russell (HR) diagram**.
- Create a page for a class digital astronomy book that successfully integrates text and images.

Materials

- Reproduction of a page from the **manuscript** *Astronomical Miscellany* (p. 10)
- Information about the featured work of art and Questions for Teaching (pp. 44–45)
- Student handout: Ursa Major's Stars (p. 48)
- Image of Ursa Major (e.g., diagram from the Montréal Planetarium website, page 10 on www.planetarium.montreal.qc.ca/Education/Fiches/PDF/constellation_an.pdf)
- Student handout: Constellation List (p. 51)
- Student handout: Constellation Research (pp. 52-53)
- Internet access or science texts for research
- Information on the Hertzsprung-Russell (HR) diagram (e.g., from the University of Nebraska-Lincoln Astronomy Education website, astro.unl.edu/naap/hr/hr_background3.html)
- White copy paper
- White or silver paper stars
- Graphics software program such as Adobe Photoshop
- (Optional) Scanner
- Microsoft Word (or another word-processing software)
- (Optional) Google Earth software, www.earth.google.com

Lesson Steps

1. Complete steps 1–4 of the intermediate-level lesson.

2. Tell students to conduct research to determine the luminosity of the stars that make up their constellations. One helpful website is How Stuff Works, science.howstuffworks.com/star3.htm.

3. Make a large Hertzsprung-Russell (HR) diagram. An HR diagram plots a star's temperature (on the x axis) against its luminosity (on the y axis). For more information, visit the University of Nebraska-Lincoln Astronomy Education website. Tell students to refer to their Constellation Research handout and their research on the luminosity of the stars that make up their constellations. Have students chart where the stars that make up their constellations are located on the HR diagram.

4. Discuss the chart by comparing the stars on the class HR diagram to an HR diagram such as that on the Science Education Gateway website, cse.ssl.berkeley.edu/segwayed/lessons/startemp/l6.htm. Do the stars fall into the "Main Sequence," or are they "Giants," "Super Giants," or "Dwarfs?"

5. Display the page from *Astronomical Miscellany*. Discuss what a contemporary version of this page would look like and discuss how computers and photographs taken with telescopes have changed scientific texts and influence our understanding of astronomy.

6. Tell students they will be using computer software to create a page for a class digital astronomy book. Complete steps 5 and 6 of the intermediate-level lesson.

7. After students are pleased with their layouts, have them re-create the star patterns of their constellations on the computer or scan in images of their constellations. Instruct them to use drawing tools to create the animals, creatures, or objects associated with their constellations and use text tools to arrange text around their constellations. Have students insert their images into a Word document.

8. Combine all of the students' individual Word documents into one class book.

Capturing Light
The Science of Photography

Students understand various properties of light through experimentation with early photographic processes. Students create and use pinhole cameras to determine how light travels and discover how artists manipulate light to create images.

Beginning Level
Students create pinhole cameras to understand that light travels in a straight path. They describe the lines and shapes in a nineteenth-century photograph of a building and then use their pinhole cameras to trace the architecture of their school building.

Intermediate Level
Students create pinhole cameras to understand that light travels in a straight path and can be refracted by a convex lens. They describe and analyze a nineteenth-century photograph of a building and then use their pinhole cameras to trace the architecture of their school or an important building in their community.

Advanced Level
Students use a pinhole camera to shoot and develop a photograph to understand how artists capture and manipulate light to make images. They compare and contrast a nineteenth-century image, photographs taken with a pinhole camera, and pictures created with a digital camera or camera phone.

Standards Addressed
Refer to the charts for national and California state standards on the Getty website, www.getty.edu/education/teachers/classroom_resources/curricula/art_science/downloads/standards.pdf

FEATURED WORK OF ART ►

Dr. John Murray
Scottish, 1809–1898
The Emperor's Private Mosque in the Marble Palace, Agra Fort, India
1857–60
Waxed-paper negative
14½ x 18 1/16 in.
J. Paul Getty Museum
98.XM.7
www.getty.edu/art/gettyguide/artObjectDetails?artobj=114911

This large waxed-paper **negative** displays a view of the private mosque built for the Mughal emperor Shah Jahan in Agra in the 1630s. Dr. John Murray, who made this negative while working in India, altered it to improve the visual harmony and luminosity of the finished positive print. He achieved increased contrast by blocking out the sky area with pigment and bleaching the deeply shaded section under the roof to more clearly show the mosque interior.

The **waxed-paper process** was particularly useful for traveling photographers like Murray because the paper did not require immediate development. It also offered more translucency than other commonly used paper negative methods. The process involved rubbing wax into the paper negative before it was sensitized and exposed. The wax created a smooth surface and reduced the blurring effects of paper fibers. Both the negative and the final print display great clarity of detail, as in the piercings of the surrounding wall and the outlines of distant buildings.

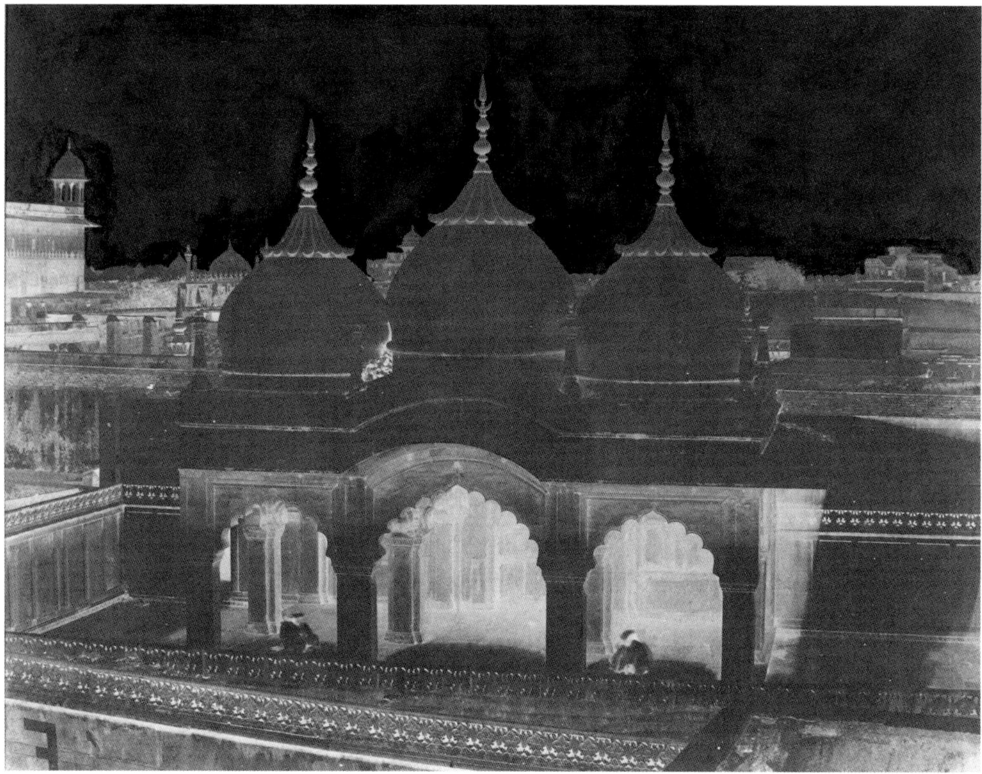

ABOUT THE ARTIST

Although trained as a medical doctor, Dr. John Murray excelled as a photographer. The Scottish-born doctor was introduced to **photography** around 1849, while in the medical service of the army of the East India Company. Stationed near the Taj Mahal in Agra, he developed a considerable interest in the Mughal architecture of the region. Throughout the forty-year period that Murray lived and worked in India, he systematically recorded many famous buildings in and around Agra and the northern state of Uttar Pradesh.

In the mid-1800s, no reliable method of enlarging **photographs** existed. To make a sizable print, Murray worked with a large-format wooden **camera** capable of accepting light-sensitive materials up to 16 x 20 inches. He employed both waxed-paper and **glass negatives**. With this unwieldy equipment, Murray produced a body of work documenting India's architecture that remained unsurpassed in the 1800s.

QUESTIONS FOR TEACHING

- What do you notice first when looking at this photograph?
- How is this picture different from pictures you take with your own camera?
- What shapes do you see?
- What lines do you see?
- Based on what you see, what could be the purpose of the building in the photograph?
- What are possible reasons for photographing this type of building?

LESSON PLAN | **BEGINNING LEVEL**

Grades: Upper elementary (3–5)
Subjects: Science and visual arts
Time required: 3 class periods

Lesson Overview

Students create pinhole **cameras** to understand that **light** travels in a straight path. They describe the **lines** and **shapes** in a nineteenth-century **photograph** of a building and then use their pinhole cameras to trace the architecture of their school building.

Learning Objectives

Students will:

- Understand that light travels in a straight path.
- Understand that the process of making photographs has changed over time.
- Identify lines and shapes in a photograph of a building.
- Draw a building by using the pinhole camera as a tracing tool.

Materials

- Reproduction of *The Emperor's Private Mosque* by Dr. John Murray (p. 11)
- Information about the featured work of art and Questions for Teaching (pp. 55–56)
- Student handout: Light Waves and Camera Pinhole Illustration (p. 59)
- Empty cardboard oatmeal canisters
- Black spray paint
- Waxed paper
- Rubber bands
- Thick black paper, cut to the width of each oatmeal canister and the length plus two inches
- Small, thin pin or needle
- Masking tape
- Journals or bound paper for recording notes
- Pencils
- Drawing paper

Lesson Steps

1. Ask students if they have seen or used digital cameras or camera phones. Tell students that taking pictures is much easier now because we simply push a button and see the picture right away. Display an image of Dr. John Murray's photograph *The Emperor's Private Mosque.* Tell students that while today's digital pictures are viewable instantly, early photographic images like Dr. Murray's had to be developed using special chemicals. Explain that this picture is a paper **negative** that was created with a camera. A negative is the image that is made with the camera, and the print is the positive image (opposite of the negative).

2. Use the Questions for Teaching to encourage students to look closely at the image. Explain that students will create a simple camera to learn about the type of basic tool that artists use to make photographs. Explain that the paper negative was created using methods similar to those they will be using in this lesson. Inform the class that light travels in a straight path and that humans learned to use light to create images on paper. Tell them that the word *photograph* means "to write with light," and that *camera* means "room." Therefore, when we take a picture with a camera, we are capturing light in a small dark room.

3. Students will create pinhole cameras. First, you will need to spray-paint black on the insides and outsides of empty cardboard oatmeal canisters. Spray paint works best for an even, thick coat of paint so that no light can come in from the sides of the canisters.
   - Give each student a pre-painted oatmeal canister, a piece of waxed paper, a rubber band, a piece of black paper, and masking tape.
   - With a pin, help students poke a hole in the center of the bottom of the canister.
   - Students will center the waxed paper over the top of the can, wrap it around the sides, and secure it with a rubber band.
   - Then students will wrap the black paper around the perimeter of the canister and align the edge of the paper with the bottom of the canister. At the top of the can, the paper should extend past the waxed paper top by at least two inches.
   - Secure the paper to the can with masking tape.

4. Take the pinhole cameras outside. Point the bottom of the canister toward the school building. Have students look into the top of the canister and cup their hands around the

edge to keep light out. Ask students the following questions:

- What do you see?
- How does the image compare to the way the building looks without the camera?
- How does the image change if you move closer to the object?
- How does it change if you move farther away? (The students should see the image on the waxed paper screen, but it will be upside down. As they move closer and farther away from the object, it will change in focus and size.)

5. Explain to students that light always travels in a straight path. When they look at an object through the simple camera, it is upside down because the light that hits the top of the object has to pass through the pinhole and will come out on the other side (the screen) in the opposite position. Have students fill in the handout Light Waves and Camera Pinhole Illustration to demonstrate the principle. Have students draw in the lines that represent the light waves by following the diagram.

6. Tell students they will capture the image of an important building, just as Dr. John Murray captured the image of the mosque. Point out that Murray created several pictures in India to document the architecture there. Tell students they will use the pinhole cameras to document the architecture of their school. Distribute pencils to students. Have students view their school building through the pinhole camera. Tell them to move around until they find an angle of the school building that they like. Allow them to trace the lines and shapes they see on the image of the building onto the waxed paper. Tell students that early cameras were often used to help people to draw. Also point out that breaking down the image into basic lines and shapes will help them to draw.

7. Instruct students to carefully remove the waxed paper from their pinhole cameras. Tell them they will use the sketch of the building as a guide to create a larger drawing of their school.

8. Display the finished drawings in a class book or on the classroom walls. Have students identify lines and shapes in their own drawings.

STUDENT HANDOUT

LIGHT WAVES AND CAMERA PINHOLE ILLUSTRATION

The lines with arrows are light waves.

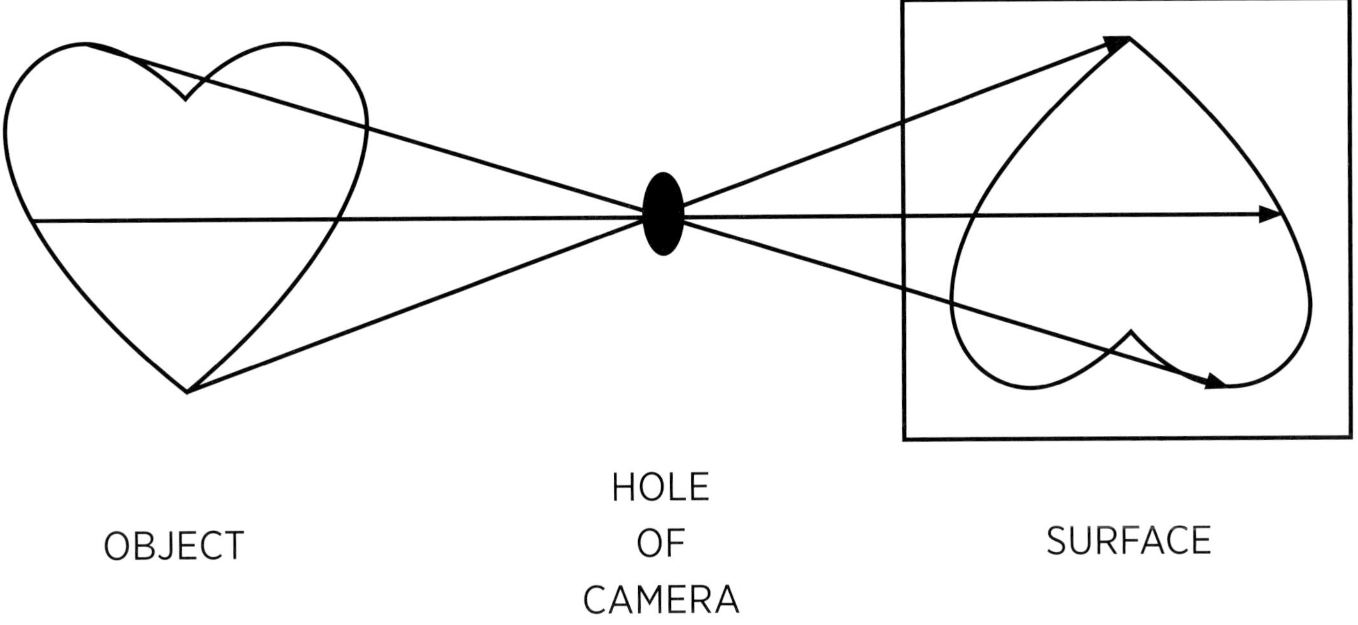

Draw your own light waves.

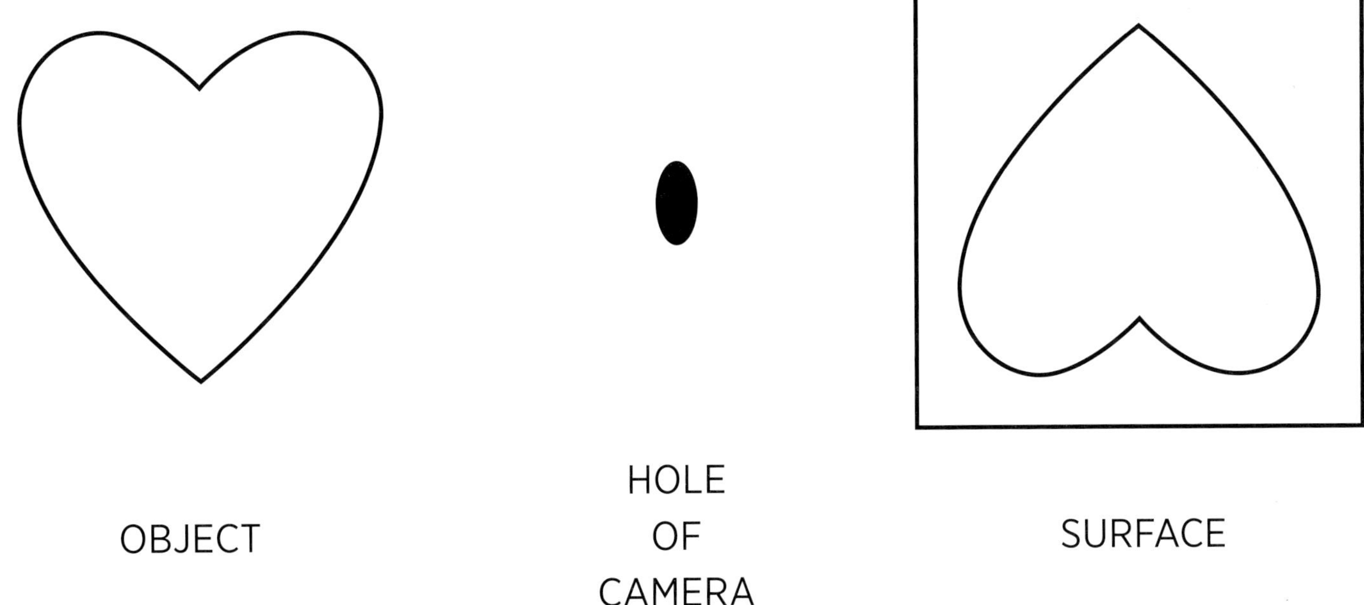

LESSON PLAN | **INTERMEDIATE LEVEL**

Grades: Middle school (6–8)
Subjects: Science and visual arts
Time required: 3 class periods

Lesson Overview

Students create pinhole **cameras** to learn how artists manipulate **light** to make **photographs**. They describe and analyze a nineteenth-century photograph and use their cameras to capture the architecture of their school or other buildings.

Learning Objectives

Students will:

- Understand that light travels in a straight path.
- Understand that the process of making photographs has changed over time.
- Draw a building by using the pinhole camera as a tracing tool.
- Understand that light can be refracted by a **convex lens** to focus a camera.

Materials

- Materials listed in the beginning-level lesson (p. 57)
- Student handout: Light Waves and Lenses (p. 61)
- Small magnifying glasses
- Reproductions of photographs printed on metal, glass, and paper. Metal: John Plumbe, Jr., *The United States Capitol,* 1846, **daguerreotype**, www.getty.edu/art/gettyguide/artObjectDetails?artobj=108465&handle=li; glass: Attributed to Henry Hollister, *Couple at Niagara Falls in Waterproofs,* 1860s, ambrotype, www.getty.edu/art/gettyguide/artObjectDetails?artobj=51451&handle=li; paper: William Henry Fox Talbot, *View of the Boulevards of Paris,* 1843, salt print, www.getty.edu/art/gettyguide/artObjectDetails?artobj=46633&handle=li.
- (Optional) Videos: *Photography: The Wet Collodion Process*, www.getty.edu/art/gettyguide/videoDetails?segid=1726), and *Early Photography: Making Daguerreotypes,* www.getty.edu/art/gettyguide/videoDetails?segid=378

Lesson Steps

1. Complete steps 1–2 of the beginning-level lesson.
2. Look at early photographs such as **cyanotypes**, **ambrotypes**, and **salted-paper prints**. Show photographs printed on metal, glass, and paper on the Getty website. Discuss how a camera was used to expose either glass, metal, or paper coated with **photosensitive chemicals** to create images.
3. Compare how images look on different materials. The images on metal and glass are sharper than the image on paper, which is slightly blurred, with less detail visible. You may have students watch the videos *Photography: The Wet Collodion Process* and *Early Photography: Making Daguerreotypes* on the Getty website to help students understand the complexity of early photographic processes. Explain that technological advances have made it easier for us to use cameras and to make **negatives** and prints.
4. Tell students they will be creating simple pinhole cameras. Point out that the basic function of a pinhole camera is similar to that of the camera used in the *Making Daguerreotypes* video. Complete steps 3–5 of the beginning-level lesson.
5. Hand out small magnifying glasses. Have students place the magnifying glasses in front of the pinholes on their cameras and look at the images on the waxed paper. Ask students to compare how their images look with and without the magnifying glasses.
6. Explain that the magnifying glass functions as a convex lens. When light passes through the **lens**, the light is refracted (bent) by the curved glass. This forces the **light waves** to move closer together as they pass through the pinhole. This function of a lens allows photographers to focus on objects from a variety of distances. When passing through a convex lens, the light waves are bent toward the optical center of the lens, causing them to converge on the focal point in front of the lens. The relative position of the object with respect to the focal point of the lens determines how the object is imaged. If the object is beyond twice the length of the focal point, then it appears smaller and inverted (see Illustration 1 on the Light Waves and Lenses handout). If the object is closer to the lens than the focal point, the image appears larger and upright (see Illustration 2 on the handout).
7. Tell students they will be using their pinhole cameras to create sketches. Complete steps 6–8 of the beginning-level lesson, or adapt the lesson by having students select and sketch an important building in their neighborhood.

STUDENT HANDOUT

LIGHT WAVES AND LENSES

ILLUSTRATION 1

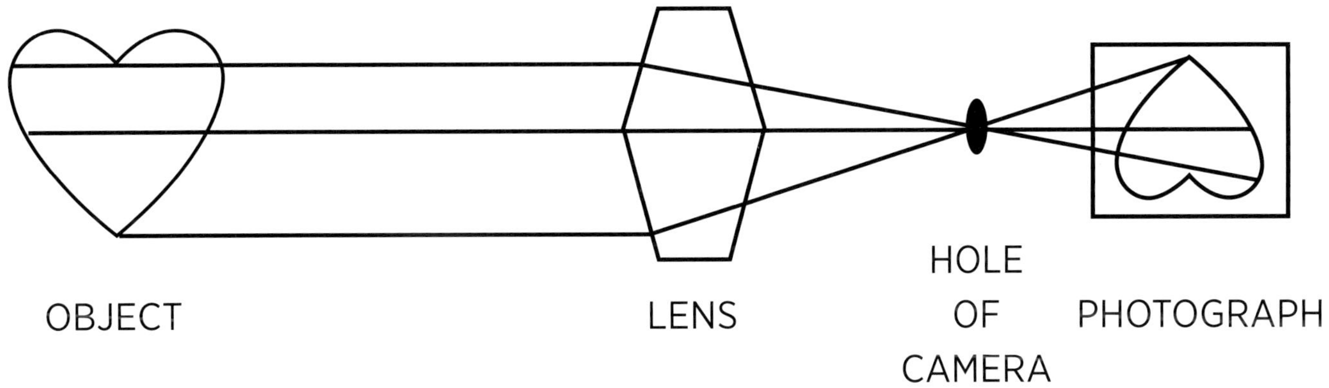

ILLUSTRATION 2

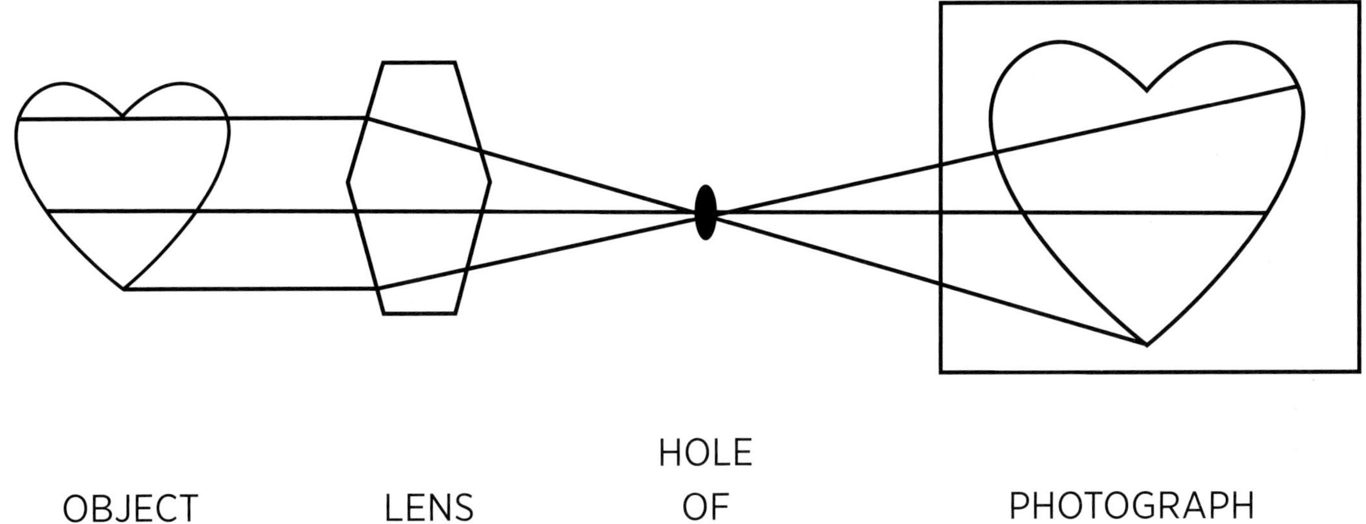

LESSON PLAN | **ADVANCED LEVEL**

Grades: High school (9–12)
Subjects: Science and visual arts
Time required: 1–3 class periods

Lesson Overview

Students create and use pinhole **cameras** to understand how artists use and manipulate **light** to capture images in **photographs**. They shoot and develop photographs made with pinhole cameras. They compare and contrast a nineteenth-century image, photographs taken with a pinhole camera, and pictures created with a digital camera or camera phone.

Learning Objectives

Students will:

- Understand that light travels in a straight path and can be refracted with a **convex lens**.
- Understand that light-sensitive chemical processes can be used to create images using light as a catalyst.
- Identify characteristics of **light waves** and make predictions about how light can be manipulated to affect a photograph.
- Create and develop photographs using a pinhole camera.
- Compare and contrast photographs made with different types of cameras.

Materials

- Materials listed in the beginning-level lesson (p. 57)
- Small magnifying glasses
- Oatmeal canister lids
- Black paint and paintbrushes
- Black electrical tape
- **Photosensitive** paper purchased at a camera supply store
- Flat objects to use as stable supports for cameras
- Examples of contemporary photographs of buildings (from books, magazines, online, or taken by the teacher with a camera, digital camera, or camera phone)
- Stationary objects like vases, books, or items from around the classroom
- (Optional) Digital camera or camera phone
- Examples of photographs printed on metal, glass, and paper listed in the intermediate-level lesson (p. 60)
- (Optional) Two videos on the Getty website listed in the intermediate-level lesson (p. 60)

Lesson Steps

1. Complete steps 1–2 of the beginning-level lesson and steps 2–6 of the intermediate-level lesson, adapting for grade level as appropriate.

2. Students will use their pinhole cameras to photograph their school building or important buildings in their neighborhood, just as Dr. John Murray photographed the mosque. Point out that Murray created several pictures in India to document the architecture there.

Camera Preparation

- Students should cover the pinhole with a piece of black electrical tape.
- They should remove the paper sleeve and waxed paper from the camera.
- Hand out canister lids.
- Have students paint the lids of the canisters with black paint.
- Set aside to dry.
- Students should gather two flat objects to place on each side of their cameras to keep them from rolling.

Paper Preparation

- Students will be given photosensitive paper to insert into their cameras, but first they should practice with regular paper that has been marked with a *B* on one side to indicate the blue side of photosensitive paper.
- Distribute regular paper marked with a *B* on one side. Tell students to place the side with the *B* against the top of the canister, and then cover the back with the lid.
- Once students can demonstrate that they can accomplish the task quickly, dim the lights in the room and distribute the photosensitive paper in its package.
- Give students about half a minute to insert the photosensitive paper in their cameras and cover with the lid. Distribute masking tape to secure the lid in place.

Exposure and Developing

- Students will walk around their school or buildings of their choice to select their **compositions**. (Optional: students can use digital cameras or camera phones to test out different compositions and angles.)
- Once they find their favorite compositions, students will put down their pinhole cameras and secure them on each side with flat objects. Students will then remove the tape from the pinhole for five minutes, replacing it when the time is up.

- Students should take their cameras to a sink and fill the sink with cold water. They will quickly remove the lid and paper from the camera and submerge the paper in the cold water for about four minutes. Remove the wet paper and allow it to dry. Students will have negative images of their subjects.

3. Students share with partners the photographs they took of buildings. Each pair should compare their **negatives** to contemporary photographs of buildings or to pictures shot with their digital cameras or camera phones. Have them discuss the following:
   - What are the differences between the images?
   - What are the things you like and dislike about the different types of photographs?
   - Which types of surfaces on the buildings created specific tones or colors in the negatives?

   Pairs should chart their comparisons and share their findings with the class.

4. Display the reproduction of the photograph by Dr. John Murray. Discuss how that negative is similar to or different from the students' photographs. Ask students to identify which areas of the mosque might have had **reflective** or dull surfaces, based on their observations of their own negative photographs.

5. Students will use their observations from steps 2 and 3 as the basis for experimenting with taking photos of two objects that have reflective and dull surfaces. Repeat the exposure process in step 2 with the objects. Hand out journals and have students record results.

6. Students will then take photographs using a **lens** with their pinhole cameras. Have students mount small magnifying glasses to their cameras with tape. Students should hypothesize about how the lens will affect their photographs, based on their knowledge of the characteristics of light waves. Repeat the exposure process in step 2. Have students record their hypotheses, observations, and findings in their journals.

7. Students can also experiment by placing an object inside the canister in front of the paper. After they expose the object and paper, they will see a silhouette of the object on the paper. Areas around the object will be fuzzy. Explain to students that the fuzzy areas were created as light waves were obstructed by and diffracted around the object.

The Magnificent Microscope

Students understand the history and evolution of the microscope as a scientific tool and designed object. They study an ornately decorated microscope from the eighteenth century to understand how it works, what it was used for, and how the tool has developed over time. Students create, decorate, and use their own microscopes based on this historical model.

Beginning Level
Students compare and contrast an eighteenth-century microscope to a modern-day class microscope. They create their own simple microscopes, decorate them with patterns, and use them to view small classroom objects.

Intermediate Level
Students use three different lenses—the human eye, a magnifying glass, and a simple compound microscope—to record their observations of hair or plant samples. They select the most effective method and create detailed drawings.

Advanced Level
Using both their decorated microscopes and a modern-day class microscope, students draw close-up views of various biological specimens and compare the results. They work in pairs to discuss the structural differences and similarities between the cells in various specimens.

Standards Addressed
Refer to the charts for national and California state standards on the Getty website, www.getty.edu/education/teachers/classroom_resources/curricula/art_science/downloads/standards.pdf

FEATURED WORK OF ART ►

Compound Microscope and Case, about 1751
Gilt bronze attributed to Jacques Caffieri
French, 1678–1755
Micrometric stage invented by Michel-Ferdinand d'Albert d'Ailly, duc de Chaulnes
French, 1714–1769
MICROSCOPE: gilt bronze, enamel, shagreen, and glass
18⅞ x 11 x 8¹⁄₁₆ in.
CASE: wood, gilded leather, brass, velvet, silver braid, and silver lace
26 x 13¾ x 10⅝ in.
J. Paul Getty Museum
86.DH.694
www.getty.edu/art/gettyguide/artObjectDetails?artobj=6789

This **compound microscope** was made for an aristocratic amateur scientist, who would have used it in his ***cabinet de curiosité*** (cabinet of curiosities) to explore the mysteries of the natural world. These *cabinets* were single rooms, or even an elaborate series of rooms, containing a variety of natural **specimens**, including shells, fossils, **minerals**, bottles of preserved animals, and stuffed exotic animals, including armadillos and crocodiles.

The Getty Museum's **microscope** still works, and the case is fitted with a drawer filled with necessary attachments, such as tweezers, extra **lenses**, and slides of such items as geranium petals, hair, fly wings, and fleas. Some of the slides are from the 1800s, indicating that the instrument was in continual use for over a century.

With attachments such as an **ocular micrometer**, the microscope incorporates the latest scientific technology of the mid-1700s. The design of its curving gilt-bronze stand was the height of the **Rococo** style when it was created. A microscope of this same model belonged to Louis XV, king of France, and was part of his observatory at the Château de la Muette.

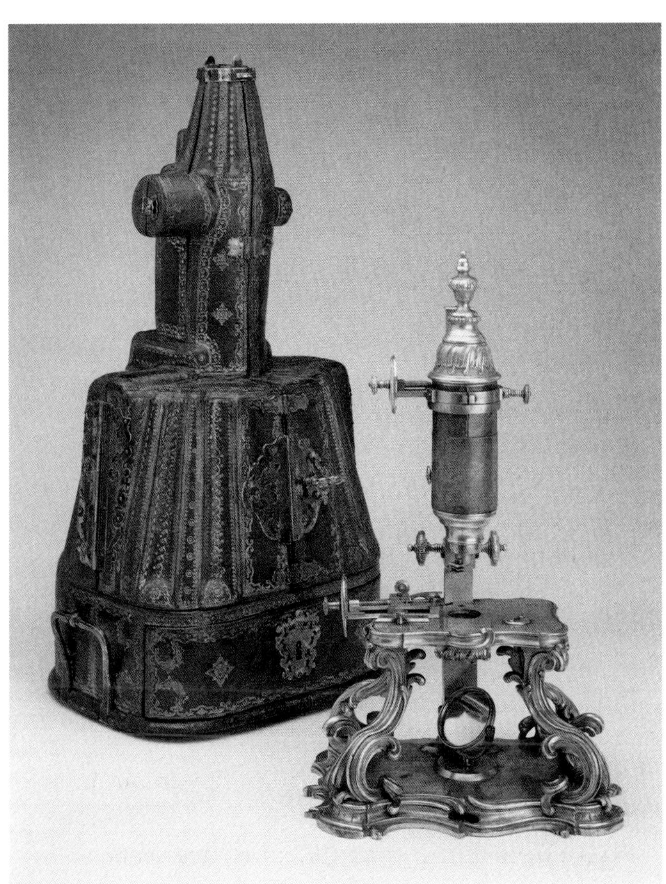

ABOUT THE ARTIST

Born into a family of sculptors and metalworkers, Jacques Caffieri became one of France's most important bronze casters during the reign of King Louis XV. As the nephew of Charles Le Brun, the chief designer and painter to Louis XIV, Caffieri had good connections as well as talent and rose quickly, becoming *sculpteur et ciseleur ordinaire des bâtiments du roi* (sculptor, bronze caster, and chaser for the king's palaces).

In 1740, Caffieri's wife bought a royal privilege—a form of permit from the king—which allowed them to gild bronze as well as **cast** it within the same workshop; these two processes would usually have been done by separate businesses. After his son Philippe Caffieri joined the workshop in 1747, they produced designs for chandeliers, ornaments for coaches, wall lights, and furniture **mounts**. Jacques was a master of the Rococo style, using elaborate curves, flowering branches, and fantastical beasts in his creations. His notable clients included the queen, Marie Leczinska; the king's mistress Madame de Pompadour; and one of the daughters of Louis XV, Madame Elisabeth.

QUESTIONS FOR TEACHING

- Describe the objects you see. What did you notice first? Why did this catch your attention?
- What colors do you see?
- What shapes do you see?
- What types of lines are used in the design?
- Which lines and shapes are repeated and form patterns?
- Imagine who would own an object like this one. Pay close attention to the case.
- Can you tell anything about the person who used this microscope by the way it looks?
- How does this device look similar to (or different from) the microscopes used by scientists today? How does it compare to a microscope you have used in science class?
- Does this microscope appear to be more decorative or functional?
- Do you consider this a work of art? What do you see that makes you say that? Does the fact that this object is functional influence your opinion?
- "Form follows function" is a principle that is related to modern architecture. Discuss this concept in relation to the microscope.

LESSON PLAN | **BEGINNING LEVEL**

Grades: Lower elementary (K–2), upper elementary (3–5)
Subjects: Science and visual arts
Time required: 3 class periods

Lesson Overview

Students study and describe an ornately decorated **microscope** from the eighteenth century to understand how it works, what it was used for, and how it compares to modern-day class microscopes. Students create their own simple microscopes, decorate them with **patterns**, and use them to view small classroom objects.

Learning Objectives

Students will:

- Understand that a microscope can be a decorative object and a scientific tool.
- Compare and contrast the design and decoration of an eighteenth-century microscope and a class microscope.
- Create and use a simple **compound microscope** to observe details of objects.
- Draw decorative patterns on their microscopes, using repeated **lines**, **shapes**, or **colors**.

Materials

- Image of an eighteenth-century compound microscope and case (p. 12)
- Information about the featured work of art and Questions for Teaching (pp. 64–65)
- Copies of information on page 64 for students
- Compound microscopes for every three students
- Student handout: Compare and Contrast Microscope Parts (p. 68)
- Information about historic microscopes, like those in "Museum of Microscopy" on the Molecular Expressions website, micro.magnet.fsu.edu/primer/museum/index.html
- Journals or bound paper
- Colored pencils
- Heavy paper, such as poster board or watercolor paper, cut into 7 x 5 inch rectangles
- Tempera paints, small containers, paintbrushes, and paper towels
- Student handout: Microscope Template (p. 69)
- Plastic mini hand magnifying glasses, with handles (about 2½ inches long, including the handle), which can be purchased from most science education supply stores
- Scissors, double-sided tape, pencils, and rulers

Lesson Steps

1. Hand out one compound microscope for every three students. Display an image of the eighteenth-century microscope by Jacques Caffieri. Have each group use the Compare and Contrast Microscope Parts student handout to examine parts on the contemporary and historical microscopes. Ask students to look at the design on each microscope and then ask the following questions:
   - What colors do you see?
   - What shapes do you see?
   - What types of lines are used in the design?
   - Which lines and shapes are repeated and form patterns?

   Have each group chart the answers on the student handout.

2. As a class, discuss the results of the students' findings. Pass out information about Caffieri's microscope and have students read it. Instruct them to add any details they missed to their charts. Identify the various parts of the class microscope as a group and invite students to add any additional similarities or differences between the microscopes' parts to their charts.

3. Look at other examples of microscopes from the past three hundred years, such as those in the "Museum of Microscopy" on the Molecular Expressions website. Explain to the class what a microscope is used for and share a brief summary of the tool's evolution, from its invention to contemporary use. Tell students where the Getty microscope fits in the timeline of the history of microscopes. Discuss how scientists in the past were very much connected to the arts and were often artists themselves. Sometimes scientists were very wealthy and wanted their instruments to be both decorative and functional.

4. Inform students that they are going to design their own microscopes. First, they will design the overall patterns that will decorate the surfaces of their microscopes and transfer the patterns onto poster board. They will then use the decorated poster board to build their microscopes. Hand out journals and colored pencils. Have students plan a decorative pattern for the surfaces of their microscopes in the journals using line, shape, and color. Once sketches are complete, have students transfer the patterns onto the

7 x 5 inch poster board or watercolor paper. Have students use tempera paints to add color to their designs. Set paints up in small containers with small water containers and paper towels so that students can clean their brushes if they need to change colors. Students should use their paintbrushes to apply small amounts of paint to their designs using thin layers of paint so that the paint doesn't smear and will dry faster. Once the decorations are completely painted, set them aside to dry.

5. Give each student a Microscope Template handout, two magnifying glasses, scissors, double-sided tape, a pencil, and a ruler. Use the following steps to demonstrate how to make a compound microscope. Then have students create their own using their decorated poster board.
   - First, cut out the template on the handout along the dotted lines. The large rectangle (A) will become the microscope. From that rectangle, cut out the long thin rectangle (B) and the small square (C).
   - Align the 7 x 5 inch template with the 7 x 5 inch poster board, with the template on top. Use the template as a stencil to draw rectangle B and square C onto the poster board.
   - Cut out the rectangle and square and remove them from the poster board.
   - Insert the handle of one of the magnifying glasses into the small square hole so that the **lens** is parallel to the bottom line of the template (**perpendicular** to the long rectangular hole). The handle should be on the decorated side of the paper and the lens on the other side. If the handle does not fit, enlarge the square with scissors. Be careful not to cut too much. The handle should fit snugly in the square hole.
   - Pull the magnifying glass's handle through the square hole until the lens touches the poster board, and wrap the poster board tightly around the edge of the lens, creating a tube out of the poster board. This tube is the microscope shaft. Make sure the decorated side of the poster board is on the outside of the tube.
   - Unroll the tube and insert the handle of the second magnifying glass into the long rectangular hole, with the handle on the same side as the other lens. Pull the handle through until the lens touches the poster board. This second lens should be parallel to and above the first one. Wrap the poster board around both lenses so that they are now inside the microscope shaft.
   - Secure the poster board around both lenses with double-sided tape, affixing tape along the shaft to hold the tube together. The handle of the second lens can now be raised and lowered above the stationary lens.

6. Once the students complete their decorated microscopes, have them look at small objects around the classroom using their microscopes and record observations in their journals. To use the microscope, place the base or bottom of the microscope over an object. Hold the handle of the bottom magnifying glass to stabilize the microscope. Hold the handle of the top magnifying glass and slide it up and down to focus. Students will share their findings with the rest of the class.

STUDENT HANDOUT

COMPARE AND CONTRAST MICROSCOPE PARTS

Compare and contrast the parts of a contemporary microscope with those of the compound microscope and case attributed to Jacques Caffieri. Look at the design on each tool. What colors do you see? What shapes do you see? What types of lines are used in the design? Record your findings below.

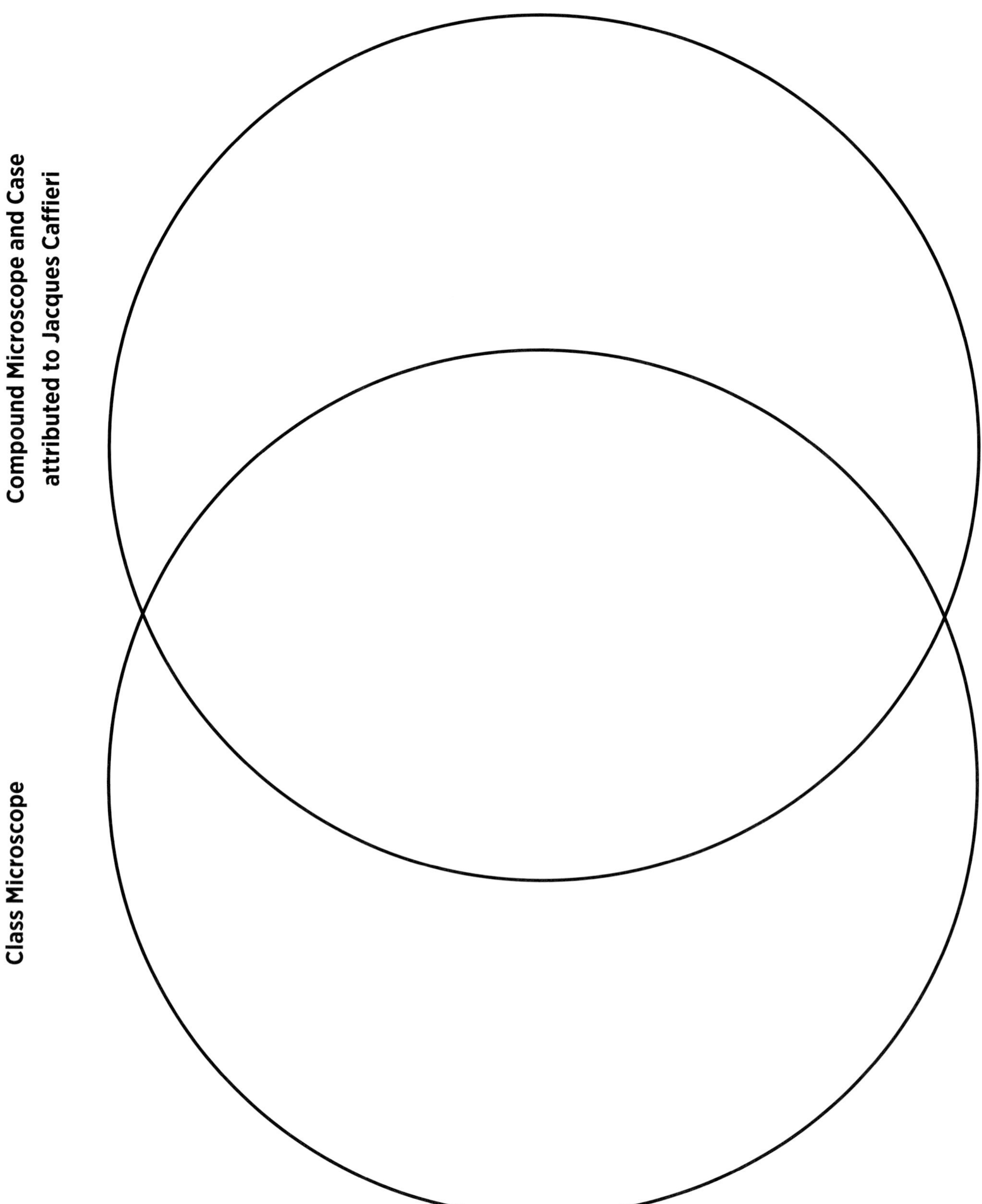

STUDENT HANDOUT

MICROSCOPE TEMPLATE

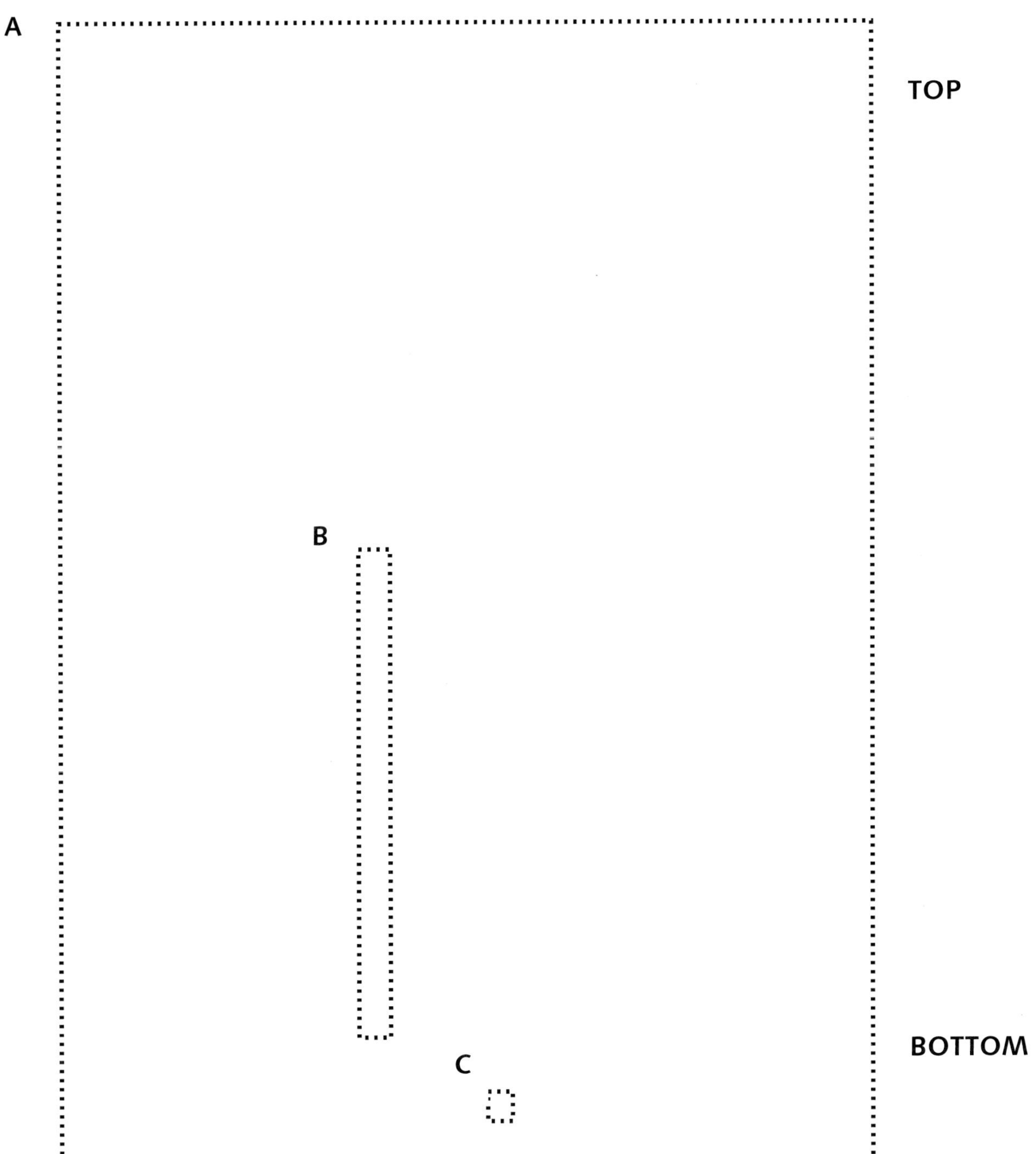

LESSON PLAN | **INTERMEDIATE LEVEL**

Grades: Middle school (6–8)
Subjects: Science and visual arts
Time required: 4–5 class periods

Lesson Overview

Students study and describe an ornately decorated **microscope** from the eighteenth century to understand how it works, what it was used for, and how it compares to modern-day class microscopes. Students create and decorate their own simple microscopes. They use three different **lenses** to record their observations of hair or plant samples, select the most effective method, and create detailed drawings.

Learning Objectives

Students will:

- Understand that a microscope can be a decorative object and a scientific tool.
- Compare and contrast the design and decoration of an eighteenth-century microscope and a class microscope.
- Understand that lenses work to focus objects.
- Observe and draw **specimens** with three different lenses—the human eye, a magnifying glass, and a **compound microscope**.

Materials

- Materials listed in the beginning-level lesson (p. 66)
- Student handout: Comparing Lenses (p. 72)
- (Optional) Digital imaging software (e.g., Adobe Photoshop)
- **Flat slides**
- Water and eyedroppers
- Art activity: "Exploring Value to Create Form," www.getty.edu/education/teachers/classroom_resources/tips_tools/downloads/exploring_value.pdf

Lesson Steps

1. Display an image of the eighteenth-century microscope by Jacques Caffieri. Allow students to share their initial observations about the object and then ask them the following questions:
   - What words would you use to describe this object?
   - What types of **lines** do you see in the design?
   - Where do you see **patterns**?
   - Imagine who would own an object like this one. Pay close attention to the case. Can you tell anything about the person who used this microscope by the way it looks?
   - How does this device look similar to (or different from) the microscopes used by scientists today?

2. Hand out one compound microscope and the Compare and Contrast Microscope Parts student handout for every three students. Complete steps 2–5 of the beginning-level lesson, adapting for grade level as appropriate. For example, students can create the designs for their microscopes using digital imaging software.

3. Have each student collect a plant sample from outside or a hair sample from his or her own head. Prepare the sample for viewing by placing a drop of water on a flat slide, laying the hair strand or plant sample in the water, and placing a second flat slide on top.

4. Distribute the Comparing Lenses student handout. Students will use pencils to draw their observations in the appropriate oval. First, they draw what they see with their naked eyes.

5. Hand out one magnifying glass per student. Students will use the magnifying glasses to observe their samples and draw what they see with pencils in the appropriate oval.

6. Students then place their slides under the microscopes they made in the beginning-level lesson, moving the top handle up and down to see how the image changes. They can also experiment with different light levels by moving to different locations in the room, such as next to a window, or by shining flashlights in the small squares on their microscopes, to see how different light levels affect what they see. Once they find a view they like, students should draw their observations with pencils in the appropriate oval.

7. Students will share their findings with the rest of the class. Discuss how simple lenses work to focus objects. Explain that the same type of lens is used in our eyes, in a magnifying glass, and in a compound microscope. Discuss

the differences they observed using the three different methods.

8. Have students select the method they think is most effective. Instruct students to write a paragraph that describes why they selected this method and then use the method to create detailed drawings of their samples using pencils and colored pencils. Have students incorporate shading techniques in the drawings as appropriate. You may wish to view the art activity "Exploring Value to Create Form" on the Getty website.

STUDENT HANDOUT

COMPARING LENSES

Collect a hair strand or a plant sample. Observe your specimen with three different lenses: your eye, a magnifying glass, and a compound microscope. Draw your observations in the appropriate oval below.

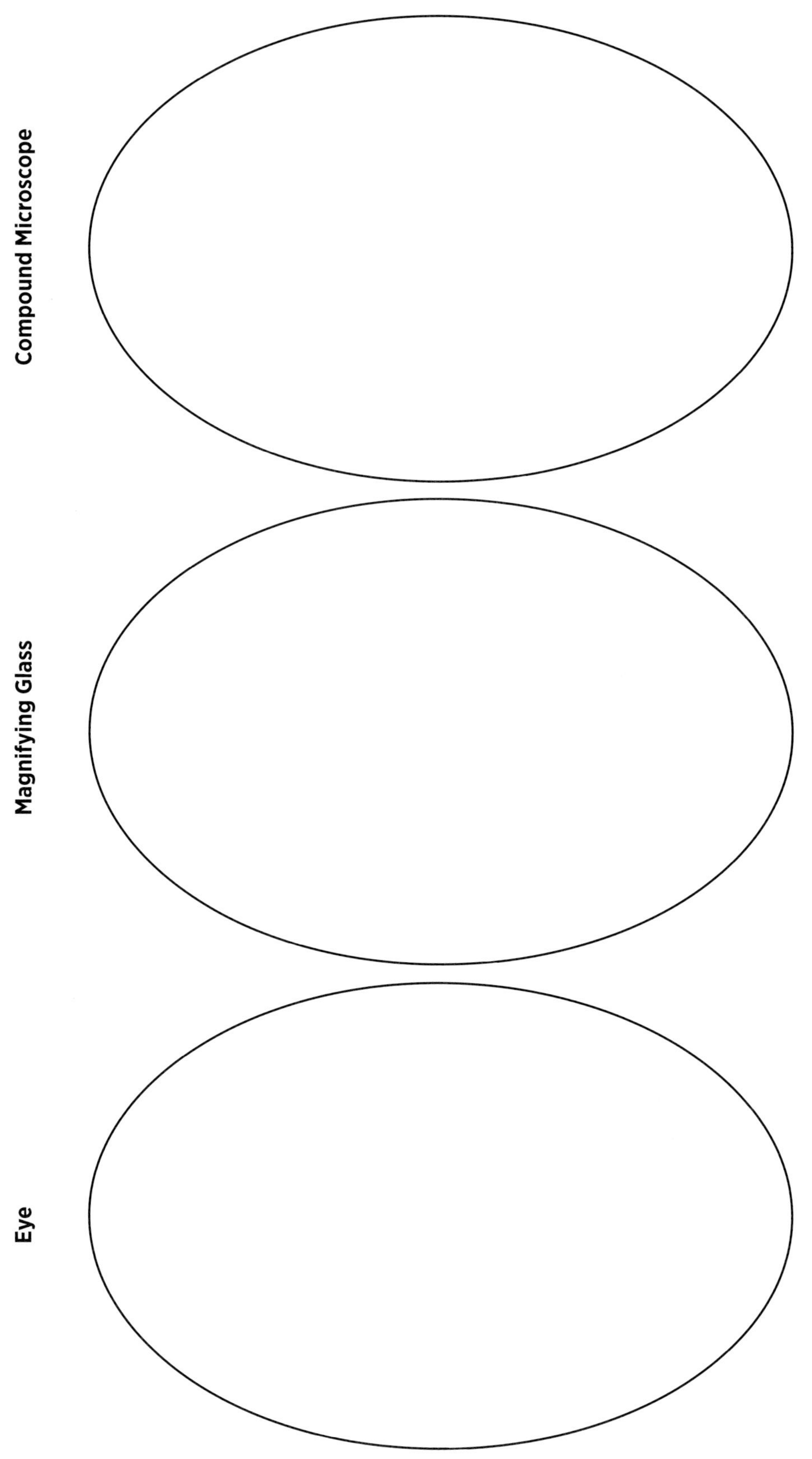

LESSON PLAN | **ADVANCED LEVEL**

Grades: High school (9–12)
Subjects: Science and visual arts
Time required: 4–5 class periods

Lesson Overview

Students study an ornately decorated **microscope** from the eighteenth century to understand how it works, what it was used for, and how the tool has developed over time. Students create their own decorated microscopes based on this historical model. Using both their decorated microscopes and modern-day class microscopes, they draw close-up views of various **biological specimens** and compare the results. Students work in pairs to discuss the structural differences and similarities between the **cells** in various specimens.

Learning Objectives

Students will:

- Understand the history and evolution of the microscope as a scientific tool and a decorative object.
- Compare and contrast the design and decoration of an eighteenth-century microscope and a class microscope.
- Design and create a simple **compound microscope**.
- Use microscopes to compare the general structure of cells.

Materials

- Materials listed in the beginning-level lesson (p. 66)
- (Optional) Digital imaging software (e.g., Adobe Photoshop)
- **Flat slides**
- Water and eyedroppers
- Student handout: Comparing Specimens
- Organic specimens: hair strands, flower petals, a flea or other insect, or an insect wing
- Bubble specimen slides
- (Optional) Art activity: "Exploring Value to Create Form," www.getty.edu/education/teachers/classroom_resources/tips_tools/downloads/exploring_value.pdf

Lesson Steps

1. Display an image of the eighteenth-century microscope by Jacques Caffieri. Allow students to share their initial observations about the object and then ask them the following questions:
   - What words would you use to describe this object?
   - What types of lines do you see in the design?
   - Where do you see patterns?
   - Imagine who would own an object like this one. Pay close attention to the case. Can you tell anything about the person who used this microscope from the way it looks?
   - How does this device look similar to (or different from) the microscopes used by scientists today?

2. Hand out one class microscope and the Compare and Contrast Microscope Parts student handout for every three students. Complete steps 2–5 of the beginning-level lesson, adapting for grade level as appropriate. For example, students can create the designs for their microscopes using digital imaging software.

3. Gather hair strands, flower petals, and insect or insect-wing samples. Insects can be ordered through various scientific supply companies or through the school district's math and science resource centers. Prepare the samples for viewing by placing a drop of water on a flat slide, laying the sample in the water, and placing a second flat slide on top.

4. Pass out the Comparing Specimens student handout. Students will use both the decorated microscopes they made and classroom microscopes to view all three specimens. They should draw the cells that they see with pencil in the appropriate circles on their worksheets. Have students incorporate shading techniques in the drawings. You may wish to view the art activity "Exploring Value to Create Form" on the Getty website. Students may also use colored pencils to add detail.

5. Students will work in pairs to discuss the structural differences and similarities between the cells in the various specimens.

6. Ask students if they can see the outline of each cell. The outline is slightly darker than the inside of the cell. Explain that this outline is a cell membrane and that all cells have **semi-permeable membranes** that keep the cells intact.

7. Ask students for examples of well-designed tools used today that can also function as objects of decorative art. Ask students to predict which of the items we use today might be collected in museums a hundred years from now for their unique designs or historical significance.

STUDENT HANDOUT

COMPARING SPECIMENS

Observe three different specimens with the compound microscope you made and with the classroom compound microscope. Draw your observations of the cells in your specimen with a pencil in the appropriate circle. Use colored pencils to add details and create form and texture in your drawings.

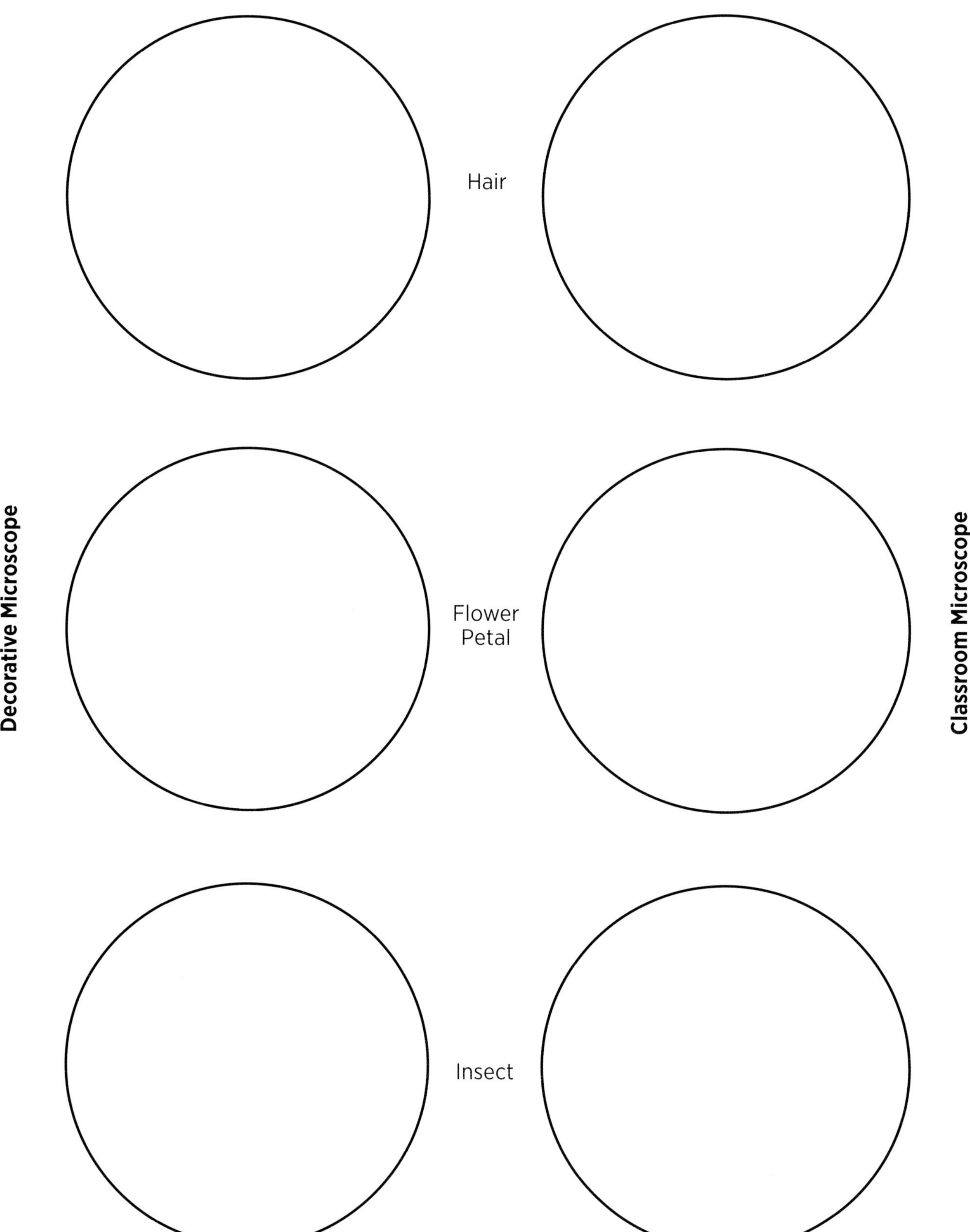

Finding Balance
Gravity, Force, and the Human Form

Students learn that artists must consider the weight of materials and the force of gravity to create balance in sculptures. They analyze the pose of a seventeenth-century figurative sculpture of a man juggling. They apply the scientific principles they learn to creating original figurative sculptures.

Beginning Level
After students analyze a sculpture of a man in a dynamic pose, they discuss how our muscles push and pull, exert force, and use strength to hold the body in a steady position. They create their own balanced sculptures out of coated wire.

Intermediate Level
Students predict how objects of different weights will affect balance. They create their own clay sculptures, applying the principles they learned about balance and gravity.

Advanced Level
Students learn about Newton's third law of gravity, centripetal force, and centrifugal force and identify action-reaction pairs in the sculpture. They create their own sculptures, applying the principles they learned about balance and gravity.

Standards Addressed
Refer to the charts for national and California state standards on the Getty website, www.getty.edu/education/teachers/classroom_resources/curricula/art_science/downloads/standards.pdf

FEATURED WORK OF ART ►

Adriaen de Vries
Dutch, about 1545–1626
Juggling Man, 1610–15
Bronze
30¼ x 20⅜ x 8⅝ in
J. Paul Getty Museum
90.SB.44
www.getty.edu/art/gettyguide/artObjectDetails?artobj=1410

At a crucial moment in an acrobatic juggling trick, this male figure holds one plate perched precariously on the fingertips of his right hand while another plate, held by **centrifugal force**, seems suspended below his left hand. Further complicating the pose, the man looks at the ground and steps on a **bellows**. The Dutch artist Adriaen de Vries based the **composition** of this **bronze** statue on a famous **Hellenistic marble** of a dancing faun, which **Michelangelo** was believed to have restored while it was in the collection of the **Medici** in Florence. Although De Vries borrowed the original statue's composition, he replaced the faun's foot organ with a bellows and substituted plates for the faun's cymbals. This sculpture combines vitality and movement with **balance**. The strong S-curve of the figure's back (see p. 76) demonstrates the complexity of his balancing act. The artist may also have had in mind the German word *kunststückemachen,* which means both to juggle and, more literally, to make a work of art. The exploration of a figure moving in space is characteristic of the **Baroque** style.

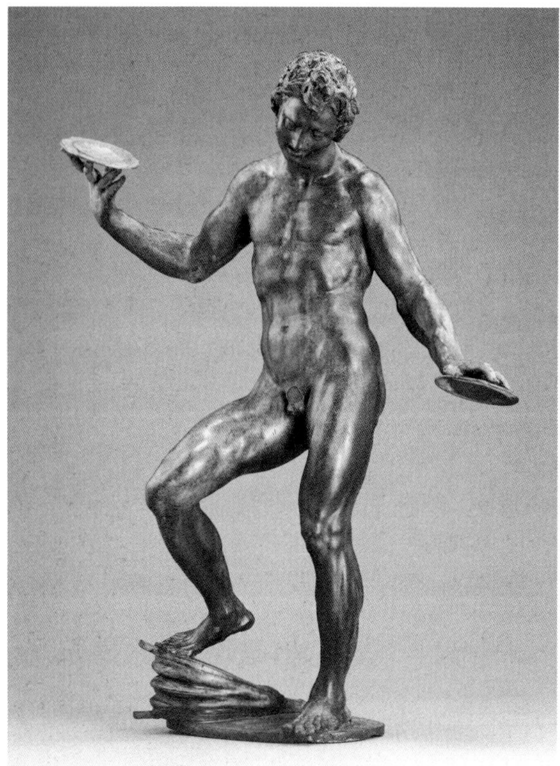

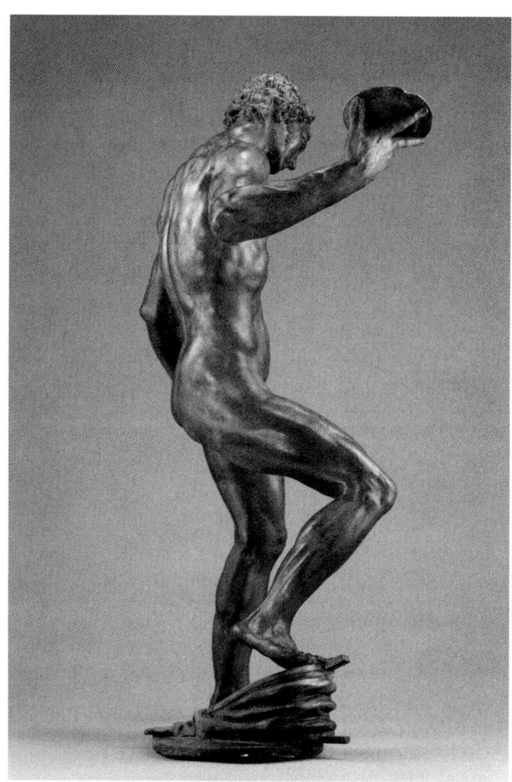

ABOUT THE ARTIST

Adriaen de Vries's career epitomizes the internationalism of late **Mannerism**. He was born in The Hague, trained in Italy, and worked mainly in Prague. His is the time-honored tradition of the itinerant artist, working for many of Europe's most discerning royal patrons. Little is known about De Vries until 1581, when he was an assistant in **Giambologna**'s Florentine workshop. There he trained as a bronze worker and absorbed much of his master's sophisticated Mannerist style. De Vries's association with Rudolf II, Holy Roman Emperor, whose works of art formed the greatest collection of the age, began in 1593. De Vries became court sculptor in 1601. Among his works for the reclusive monarch was a bronze relief representing Rudolf II's 1585 imperial decree that painting should be considered among the liberal arts. The idea that visual artists should be raised above the level of craftsmen developed during the Italian Renaissance, but Rudolf II made it official. After Rudolf II's death in 1612, De Vries continued working for aristocratic clients, creating numerous funerary monuments, life-size sculptures, fountains, and church fonts. In his late style, he tended toward soft, sketchy effects in his sculptural surfaces.

QUESTIONS FOR TEACHING

- What is going on in this artwork?
- Notice the plates and how they are held in the hands. What do you think the figure might be trying to do?
- Try out the pose of the figure. What type of movement or action is the artist trying to capture?
- What do you think this sculpture is made of? What do you see that makes you say that?

LESSON PLAN | **BEGINNING LEVEL**

Grades: Lower elementary (K–2), upper elementary (3–5)
Subjects: Science and visual arts
Time required: 2 class periods

Lesson Overview

Students learn that an artist must consider the weight of material and the **force** of gravity to create **balance** in a sculpture. They analyze the dynamic pose of a figure in a seventeenth century sculpture. They discuss how our muscles **push** and **pull,** exert force, and use strength to hold the body in a steady position. They create their own balanced sculptures out of coated wire.

Learning Objectives

Students will:

- Imitate the pose of the statue to understand kinesthetically force and balance.
- Understand that the amount of force required to balance an object is directly related to its weight.
- Analyze and interpret the pose of a figurative sculpture.
- Apply what they learn about balance to creating a figurative sculpture in a dynamic pose.

Materials

- Image of *Juggling Man* by Adriaen de Vries (pp. 13–14)
- Information about the featured work of art and Questions for Teaching (pp. 75–76)
- Journals or bound paper for recording observations
- Thin coated wire (e.g., Twisteezwire)
- Pieces of cardboard (one per student)
- Stapler

Lesson Steps

1. Display the image of Adriaen de Vries's *Juggling Man.* Give students one to two minutes to examine it carefully. Have students share their initial observations and then ask the following questions:
   - What is going on in this artwork?
   - Notice the plates and how they are held in the hands. What do you think the figure might be trying to do?
   - Try out the pose of the figure. What type of movement or action is the artist trying to capture?

2. Pass out journals and instruct students to draw the outline of the figure. This will help them to look even closer at the artwork. Introduce the term *balance* to the class. Ask students to hypothesize about whether the statue is balanced and record their hypotheses next to their drawings.

3. Divide the class into pairs. One student will imitate the pose of the figure in the sculpture. The other student should double-check the pose and help correct it, if necessary. He or she will adjust the other's body until it is as close to the *Juggling Man*'s pose as possible. Ask the students in the pose to describe to their partners how it feels to stand on one leg with their arms in the air. Ask students to describe which muscles feel like they are being worked the most in this pose. Which direction do their bodies lean to help them hold the pose? If they lean in the opposite direction, what happens? Students should record their answers in their journals.

4. Introduce the terms *push, pull,* and *force.* Ask one student to take the juggler's pose in front of the class. Explain to the class, using the student as an example, how our muscles push and pull to exert force and use strength to hold the pose and keep the body from moving in other directions. Apply force to the student's hand by pushing on one of the hands being held in the air. As you do this, ask the student to use his or her muscles to push back against you while keeping his or her arm in the same place. Ask the student to describe to the class what he or she has to do in order to hold the pose while resisting the force being placed on the hand. Do this exercise again, but this time instruct the student not to push back, and see what happens. How is the arm affected? How does the pose change? Finally, pull on the leg the student is holding up. Have the student describe what he or she needs to do to maintain the pose. Students should record their observations in their journals.

5. Display the view of *Juggling Man* showing the S-curve of the figure's back (p. 76). Tell students that the artist has arranged the body into an S-curve to combat gravity by balancing force to keep the figure upright. Share background information about the artwork with the class. Based on their experimentation with poses and findings from the discussion, students should be able to determine which parts of the statue are exerting force to balance a push or a pull.

6. Distribute thin, coated wire to each student. Have students twist and bend the wire to create sculptures of figures performing actions. You may wish to have students try out their poses for each other and observe how minor changes in position affect stability.

7. Have students problem-solve until their figures can stand on their own. Students may need to make adjustments to the angles of the legs and arms of their figures so they can stay balanced. Once the figure stays relatively balanced, have students staple their sculptures to a piece of cardboard and continue to make further adjustments as necessary.

LESSON PLAN | **INTERMEDIATE LEVEL**

Grades: Middle school (6–8)
Subjects: Science and visual arts
Time required: 3 class periods

Lesson Overview

Students learn that artists must consider the weight of material and the **force** of gravity to create **balance** in sculptures. Students predict how objects of different weights can affect balance. They create their own **clay** sculptures, applying the principles they learned about balance and gravity.

Learning Objectives

Students will:

- Imitate the pose of the statue to understand kinesthetically force and balance.
- Understand that the amount of force required to balance an object is directly related to its weight.
- Apply what they learn about balance and force to creating figurative sculptures in dynamic poses.

Materials

- Image of *Juggling Man* by Adriaen de Vries (pp. 13–14)
- Information about the featured work of art and Questions for Teaching (pp. 75–76)
- Journals or bound paper for recording observations
- Variety of objects of different weights that can fit in students' hands
- Scales for weighing objects
- Student handout: Force and Balance Observations (p. 80)
- Pencils
- Modeling clay
- Plastic utensils and popsicle sticks for sculpting clay
- Video: *Casting Bronzes,* www.getty.edu/art/gettyguide/videoDetails?segid=370

Lesson Steps

1. Complete steps 1–5 of the beginning-level lesson.

2. Divide the class into groups of three. Explain that students will explore how objects of different weights can affect balance. They will predict how a balanced pose will need to change in order to accommodate objects of differing weights. One student will take the pose of *Juggling Man.* Another student will arrange the body to make sure it is in the correct position and will then place objects of varying weight in the hands of the student holding the pose. Using the Force and Balance Observations student handout, the third student will document the objects used and the effects of adding weight. Instruct the students to perform the experiment twice for each object. The first time, the posing student should not resist the weight of the object. What direction does the pose fall? Students should then predict what the posing student will have to do to adjust the balance. When the object is placed in the posing student's hand a second time, the student should adjust his or her pose to stay balanced. The third student will record the adjustment in force that was required to keep balance. Does this reaction match the students' predictions?

3. Demonstrate how Adriaen de Vries created his bronze by showing the video *Casting Bronzes* on the Getty website. Explain that bronze has a property called **tensile strength** due to its density. By taking X-rays, conservators found that some areas of the statue are thicker than others but still remain balanced, like the solid **metal** hands resting on hollow arms. Both the tensile strength of bronze and the thicker areas in the sculpture counteract the force of gravity, helping to keep the sculpture balanced and preventing it from breaking from the weight of the metal.

4. Have students sculpt their own balanced figures of athletes in motion. Gather various pictures of people in dynamic poses, such as a baseball player swinging a bat, a soccer player kicking a ball, or a dancer doing a turn. Images can be found in magazines or newspapers. Students may also bring in photos from home. Students will choose one image from which to create a sculpture using modeling clay. Students should give special attention to the direction of body parts and how they compensate for the force being applied to the body. Students may use plastic utensils and popsicle sticks to sculpt the clay. They may also make some areas thinner and others more solid to accommodate the forces of weight and gravity. Once completed, sculptures should be freestanding and balanced.

5. Look closely at the sculptures. Have students point out which body parts are creating a force.

STUDENT HANDOUT

FORCE AND BALANCE OBSERVATIONS

Explore how objects of different weights can affect balance. Use this chart to record your predictions and observations about how a balanced pose changes to accommodate different objects.

1. Record each object and its weight, using a scale, in the first column.
2. Have one person in your group take the pose of Juggling Man. Another person will arrange the body to make sure it is in the correct position.
3. Place the object in the posing student's hand. Record the direction of the force from the object in the second column. Then revove the object from the student's hand.
4. Predict what the posing student will need to do to rebalance the pose and record it in the third column.
5. Put the object back into the student's hand and observe how the body adjusts in order to keep the pose in balance. Record your findings in the last column.

Object and weight	Change of direction	Predicted method in which change in direction will be balanced	Observed method in which change in direction was balanced

LESSON PLAN | **ADVANCED LEVEL**

Grades: High school (9–12)
Subjects: Science and visual arts
Time required: 1–3 class periods

Lesson Overview

Students discover that artists must consider the weight of material and the **force** of gravity to create **balance** in sculptures. They learn about **Newton's third law of gravity**, **centripetal force**, and **centrifugal force** and identify action-reaction pairs in a sculpture. They create their own sculptures, applying the principles they learned about balance and gravity.

Learning Objectives

Students will:

- Imitate the pose of the statue to understand kinesthetically force and balance.
- Understand that the amount of force required to balance an object is directly related to its weight.
- Apply what they learn about balance and force to creating figurative sculptures in dynamic poses.
- Understand Newton's third law of gravity as it applies to the creation of sculpture.

Materials

- Materials listed in the intermediate-level lesson (p. 79)
- (Optional) Lightweight balls—two for each pair of students

Lesson Steps

1. Complete steps 1–5 of the beginning-level lesson and steps 2–3 of the intermediate-level lesson as appropriate to grade level.

2. Discuss Newton's third law of gravity: "For every action, there is an equal and opposite reaction." Ask students to describe what this statement means. After students discuss, explain that there are always two forces acting in opposition that keep an object stable. The two forces on the object are equal—the direction of one force on the object is moving in the opposite direction of the other. Forces always come in pairs of equal and opposite—action and reaction. For instance, if you drop a ball, gravity pulls the object down toward the earth. When it hits the ground, the force of the earth pushes it back up to create the bounce.

3. Introduce the terms *centripetal force* and *centrifugal force*. Explain that De Vries depicted a man juggling. Juggling is an example of an action-reaction pair. If the sculpture were to come to life, the man's hands would push on the plates to create the centripetal force that would send the plates along a circular path toward the center. The opposite and equal reaction, centrifugal force, would be placed on the source of the force, the hands, displacing them outward from the center of the circular path. If the juggler stopped moving his arms, the initial force would be removed. The plates would continue to move in straight paths as described by Newton's first law of gravity.

4. Look at the artwork again. Ask students to describe how the man is holding the plates in each hand. Ask them to describe the shape of his arms. Students should notice that his arms are curved and that the plates are barely touching his hands, showing that the artist is capturing a snapshot of the man juggling the plates. Students should identify the action-reaction pairs in the artwork. If students have trouble understanding the centripetal/centrifugal concept, show the helpful animations on the Physics Classroom website, www.physicsclassroom.com/mmedia/circmot/cf.html.

5. Tell students they will create their own figurative sculptures using clay. Complete steps 4–5 of the intermediate-level lesson. Have students point out which body parts are creating a force. Ask students where they see the reaction to the initial force. Students may work with a partner to identify where the artists have depicted action-reaction pairs. Students may take the pose of each sculpture and mimic the action depicted. As they move, they should pay close attention to where they feel a push or a pull and the reaction that happens to keep them in balance.

Extension

Divide the class into pairs and give each pair two small, lightweight balls. One student will juggle balls while the other observes how changes of speed and motion of the arms affect the juggling. Students will switch roles. Students will record their observations and conclusions in their journals.

Medieval Natural Resources

Students study and analyze a page from a fifteenth-century manuscript. They identify and research the natural resources that were used to make it. They make paint and then use it to create their own illuminations.

Beginning Level
Students identify the resources used to create a fifteenth-century manuscript that are still used to make books today. Students create books using natural resources or found materials and make paint they can use to illustrate their books.

Intermediate Level
Students research the materials used to make manuscript pages and determine which are renewable resources. They create books using natural resources or found materials, and they make paint they will use to illustrate examples of renewable and nonrenewable resources in their handmade books.

Advanced Level
Students research the materials used to create a fifteenth-century manuscript page, determine which are renewable resources, and identify those found or produced in their state. Students make paint they will use to illustrate resource maps of their state, showing the locations in the state where each manuscript resource can be found.

Standards Addressed
Refer to the charts for national and California state standards on the Getty website, www.getty.edu/education/teachers/classroom_resources/curricula/art_science/downloads/standards.pdf

FEATURED WORK OF ART ►

The French King at Court
From the manuscript *The Story of Two Lovers*
French, 1460–70
Tempera colors and gold paint on parchment
6 15/16 x 4½ in.
J. Paul Getty Museum
Ms. 68, fol. 1
www.getty.edu/art/gettyguide/artObjectDetails?artobj=143858

The story in this image is told in a continuous narrative—characters are shown more than once in different parts of the image performing different actions. This method allowed the artist to tell a full story in one scene. A fashionably dressed young man approaches the king of France, who sits on a throne draped in a blue fabric patterned with the French **fleur-de-lis**. The courtier, perhaps a messenger, wears long riding boots and spurs. This suggests that he may be the same figure as the black-clad rider on the white horse visible in the **miniature**'s background. Both the ceremony and the intrigue of court are represented in the subtle gestures of the figures and the discrete groups of attendants around the king. The king peers out of the miniature with a seemingly knowing expression, as if to indicate his wariness of competing groups. The miniature sets the tone for the text, which concerns the perils of life at court.

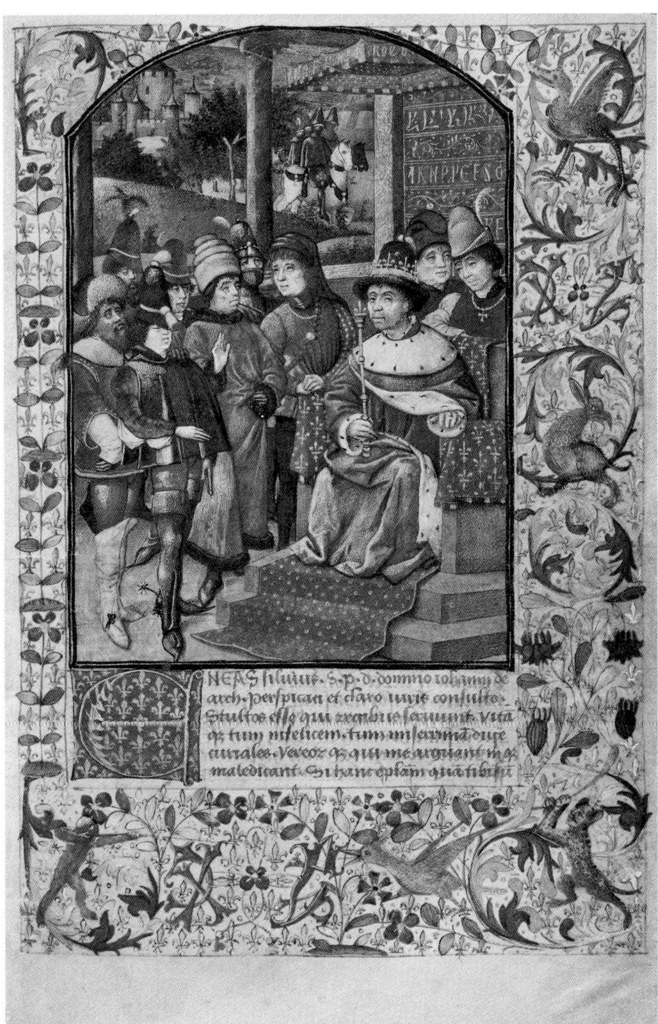

MATERIALS USED IN MANUSCRIPTS

Materials used in making **manuscript** pages come from many different sources. This page of *The Story of Two Lovers* is parchment, which is made from the skin of goats, cows, or sheep. The **lines** on which the text was written are lead, graphite, or silver. The ink for the text was made of charcoal and pine pitch. Ground **pigments** of **lapis lazuli** (bright blue), madder (red), parsley (green), saffron (yellow), and **vermilion** (bright red) were mixed with egg to create the tempera paint. Ground gold was also mixed with an egg base to create the gold paint used throughout the image.

QUESTIONS FOR TEACHING

- What do you notice about the characters in the manuscript?
- What does their clothing tell you about who they are?
- Who do you think are the main characters in this story? What do you see that makes you say so?
- What do you notice about the setting? What do you see in the foreground? What do you see in the background?
- Notice that one character is repeated in the foreground and background. Which character is it?
- What do you think is happening in the image? What do you see that makes you say that?
- Where do you see other objects repeated in the image? What kinds of patterns do they create?
- What colors do you see?
- Notice the border decoration. What images do you see in it?

LESSON PLAN | **BEGINNING LEVEL**

Grades: Lower elementary (K–2), upper elementary (3–5)
Subjects: Science and visual arts
Time required: 3 class periods

Lesson Overview

Students speculate about all of the natural resources that were used to make illuminated **manuscripts** in the Middle Ages. They determine which of these resources are still in use today. Students create books using natural resources or found materials and make paint they will use to illustrate their books.

Learning Objectives

Students will:

- Identify the natural resources used to create an illuminated manuscript page.
- Speculate about why the resources used to create books have changed over time.
- Create handmade books using natural resources or found materials.
- Make paint by crushing a variety of **pigments** and mixing pigments with a **binder**.

Materials

- Reproduction of the page depicting *The French King at Court* from the manuscript *The Story of Two Lovers* (p. 15)
- Information about the featured work of art and Questions for Teaching (pp. 82–83)
- Copy of the book *Marguerite Makes a Book* by Bruce Robertson, available from the Getty bookstore, shop.getty.edu/products/marguerite-makes-a-book-978-0892363728
- Video: *Making Manuscripts*, www.getty.edu/art/gettyguide/videoDetails?segid=372
- Information from the exhibition *The Making of a Medieval Book*, www.getty.edu/art/exhibitions/making/
- Samples of parchment
- (Optional) Samples of **lapis lazuli** stone, cochineal beetles, and gold leaf, purchased from an art supply store
- Natural and found materials to create handmade books (e.g., bark, twigs, leaves), View the following links for book-making ideas and related supplies:
  "Binding on a Budget" on the Bookbinding Etsy Street Team blog, www.bookbindingteam.com/2010/05/binding-on-budget-found-materials.html
  "Natural Twig Journals" on the Dick Blick website, www.dickblick.com/lesson-plans/natural-twig-journals/
  "Save Trees by Making Your Own Recycled Paper" on the EcoKids website, www.ecokids.ca/pub/fun_n_games/printables/activities/assets/science_nature/paper_making.pdf
- Rulers or pieces of cardboard rectangles for creating borders
- Art activity: "Making Paint," www.getty.edu/education/teachers/classroom_resources/tips_tools/downloads/making_paint.pdf
- Gum arabic, raw ochre, **raw umber**, and ultramarine powdered pigment (alternatively, use ground-up pieces of colored chalk, powdered-drink mix, instant coffee, or tea)

Lesson Steps

1. To motivate students and provide context about illuminated manuscripts and how they were made, read *Marguerite Makes a Book* to the class.

2. Display a reproduction of the manuscript page *The French King at Court* or hand out photocopies. Use the Questions for Teaching as appropriate to lead a discussion about the work of art. Chart students' responses on the board. Share background information about the page with the class. Explain that at the time the book was made, it had to be done entirely by hand using resources found in nature. Students may also watch the video *Making Manuscripts* and view information from the exhibition *The Making of a Medieval Book* on the Getty website.

3. Have students speculate about the materials that would have been used to make this page from the manuscript *The Story of Two Lovers*. Chart the materials mentioned, providing guidance as needed. (See the Materials Used in Manuscripts section on page 83.)

4. Pass out samples of parchment and have students look closely to see the pores in the skin. You can show students any other resources used to make illuminated manuscripts, including substances used for pigments. If available, hand out pieces of lapis lazuli stone for students to handle, and pass around the following fragile materials in glass or plastic vials: ultramarine, raw umber, cochineal beetles, and gold leaf.

5. Display *Marguerite Makes a Book.* Ask students to speculate about the resources that went into making this

contemporary book. Chart this list of resources next to the resources students identified for the manuscript. These materials include paper made from trees and man-made inks. Ask students to speculate about why the materials for bookmaking have changed. One source of information on this subject is the page about book production on the ORB (Online Reference for Medieval Book Studies) website, www.the-orb.net.

6. Choose one of the raw materials used to make a medieval manuscript and have students discuss with a partner how they think that resource is turned into paint, parchment, ink, etc. Students should write down their speculations or draw three or four key steps in the process they discussed.

7. Have students make books out of local natural resources such as leaves and bark. Limit the materials to resources students can obtain without spending money; this will reinforce the concept that art making is sometimes limited by economics, resources, and geography. For ideas on how to create books using found materials, visit the Bookbinding Etsy Street Team blog. For a lesson on how to use a twig to bind a book, visit the Dick Blick website. For instructions on creating paper from recycled material, visit the EcoKids website.

8. Return to the image of the fifteenth-century manuscript page. Ask students to recall what they remember about the story depicted on the page. Refer back to the chart created in step 2 to reinforce students' responses. Ask students what they think happens next in the story depicted. Tell them they will paint what they think happens next using paints made by hand, like medieval painters would have done.

9. Hand out rulers or precut rectangles to help students draw large rectangles on the pages of their handmade books. Have students draw what happens next inside their rectangles. In the borders outsides the rectangles, students should draw **patterns** inspired by *The French King at Court.*

10. Instruct students to make paint using the "Making Paint" art activity on the Getty website. If powdered pigments are not available, you may substitute ground-up pieces of colored chalk, powdered-drink mix, instant coffee, or tea. Once students have completed making paint, have them paint in their handmade books, what they think happens next to the French king.

LESSON PLAN | **INTERMEDIATE LEVEL**

Grades: Middle school (6–8)
Subjects: Science and visual arts
Time required: 3–4 class periods

Lesson Overview

Students identify the natural resources used to make illuminated **manuscripts** during the Middle Ages. They research these materials to determine which are renewable resources and then create books using natural resources or found materials. They make paint they will use to illustrate renewable and nonrenewable resources in their handmade books.

Learning Objectives

Students will:

- Research and identify the natural resources used to create manuscript pages.
- Classify the resources used to make manuscripts as renewable or nonrenewable.
- Create handmade books using natural resources or found materials.
- Make paint by crushing a variety of **pigments** and mixing the pigments with a **binder**.

Materials

- Materials listed in the beginning-level lesson (p. 84)

Lesson Steps

1. Complete steps 2–4 of the beginning-level lesson, adapting for grade level as appropriate.

2. Have students work with partners or in small groups to determine where the resources they charted in step 1 are found in the natural world. Students should then determine which resources are renewable and which are nonrenewable. Use the information on page 83 to support discussion.

3. Have students share their findings. Ask students if they think books today contain more renewable resources than manuscripts from the Middle Ages. Are contemporary books more friendly to the environment than medieval books? Why or why not?

4. Complete steps 7 and 9 of the beginning-level lesson. Have students use the paint they make to illustrate one renewable resource and one nonrenewable resource in their handmade books.

LESSON PLAN | **ADVANCED LEVEL**

Grades: High school (9–12)
Subjects: Science and visual arts
Time required: 3–4 class periods

Lesson Overview

Students research the materials used to create a fifteenth-century **manuscript** page, determine which are renewable resources, and identify those found or produced today in their state. They make paint they will use to illustrate a resource map of their state, showing the locations where each manuscript resource comes from.

Learning Objectives

Students will:

- Identify and research the natural resources used to create medieval manuscripts.
- Classify these resources as renewable or nonrenewable.
- Determine which of these resources are found or produced in their state.
- Make paint by crushing a variety of **pigments** and mixing the pigments with a **binder**.
- Paint an illuminated resource map of their state.

Materials

- Materials listed in the beginning-level lesson (p. 84)
- Assorted locally sourced materials that can be used as pigments (e.g., beets, blueberries, spinach) and binder (local eggs)
- Internet access or earth science texts for research
- One color image per student of the featured work of art (p. 15)
- Poster board
- Pens or thin markers
- Information from the exhibition *French Manuscript Illumination of the Middle Ages*, www.getty.edu/art/exhibitions/french_mss/

Lesson Steps

1. Complete steps 2–4 of the beginning-level lesson and steps 2–3 of the intermediate-level lesson, adapting for grade level as necessary.

2. Working in pairs, students will make lists of the resources researched in the intermediate-level lesson. They will identify through further research those that are native to their state, and they will determine where in their state these resources can be found. If students find that some materials used to make medieval manuscripts are not native to their state, they should identify substitutes that could fulfill the same functions.

3. Mount the images of the featured work of art onto larger pieces of paper or poster board. Give partners the opportunity to label the elements of the manuscript page that they would use native resources to re-create.

4. Have students work in pairs to make illuminated resource maps of their state showing the locations where each manuscript resource originates. Have students read information about medieval manuscripts in the overview of the exhibition *French Manuscript Illumination of the Middle Ages* on the Getty website. You may wish to print copies of this page in advance and hand them out to students.

5. Instruct students to make paint using the "Making Paint" art activity on the Getty website. Challenge students to adapt the art activity by using local resources to create the paint (e.g., regionally grown beets, blueberries, spinach, and eggs, or even dirt and burned twigs).

6. In their illuminated maps, students should follow the techniques of manuscript **illumination**. Students can look at other manuscripts on the Getty website to find authentic details they may wish to incorporate into their maps.

Fire, Water, and Earth
The Chemistry of Ceramics

Students study porcelain sculptures to understand how heat can transform clay into a hard ceramic. They sculpt decorative art objects and allow them to dry, and they discover the artistic and scientific processes involved in using clay to create sculptures. Students discuss the textures and motifs on a set of sculptures that may have been part of a candelabrum and apply what they learn from making their own ceramic sculptures.

Beginning Level
Students identify natural motifs, shapes, and textures in decorative art objects and then sculpt clay to create their own objects. They make observations about how a clay's mixture affects its properties.

Intermediate Level
After students complete the beginning-level lesson, they learn about the processes of radiation, convection, and evaporation through the drying and firing of clay.

Advanced Level
After completing the beginning-level and intermediate-level lessons, students understand the concepts of vitrification, radiation, and convection and discuss how minerals in clay and glaze absorb heat when melting.

Standards Addressed
Refer to the charts for national and California state standards on the Getty website, www.getty.edu/education/teachers/classroom_resources/curricula/art_science/downloads/standards.pdf

FEATURED WORK OF ART ►

Ginori Porcelain Factory
Italian, 1735–present
Mercury and Argus and *Perseus and Medusa*, 1749
After models by Giovanni Battista Foggini
Italian, 1652–1725
Porcelain, polychrome enamel, and parcel gilt
Each: 17¾ x 13 x 11 in.
J. Paul Getty Museum
94.SE.76
www.getty.edu/art/gettyguide/artObjectDetails?artobj=1478

Two brightly colored figural groups from the Ginori Porcelain Factory depict violent episodes from the Roman poet **Ovid**'s ***Metamorphoses***. Modeled in **porcelain**, Jupiter's messenger Mercury attacks Argus on one pedestal, while Perseus kills Medusa on another. The factory reused Giovanni Battista Foggini's bronze-casting molds, replicating the **Baroque** sculptor's muscular figures, wind-blown dramatic draperies, and dynamic **compositions**. Yet the painting, using jewel-like yellow, purple, blue, and pink **enamel pigments** and **gilding**, brings out a delicacy more typical of the **Rococo** style of the mid-1700s. As a result, Medusa, whose horrifying look was reputed to turn men to stone, is merely a rosy-cheeked woman with snakes for hair, while Argus, the hundred-eyed giant who tormented Jupiter's lover Io, is a handsome, muscled warrior.

Each figural group is made from ten separate sections. These were **press-molded** and fired before being joined together. The figures are joined to Rococo bases with candle sockets, indicating that they were meant to function as part of a **candelabrum**. The two lively groups would have been used as a dramatic centerpiece for a dinner-table setting.

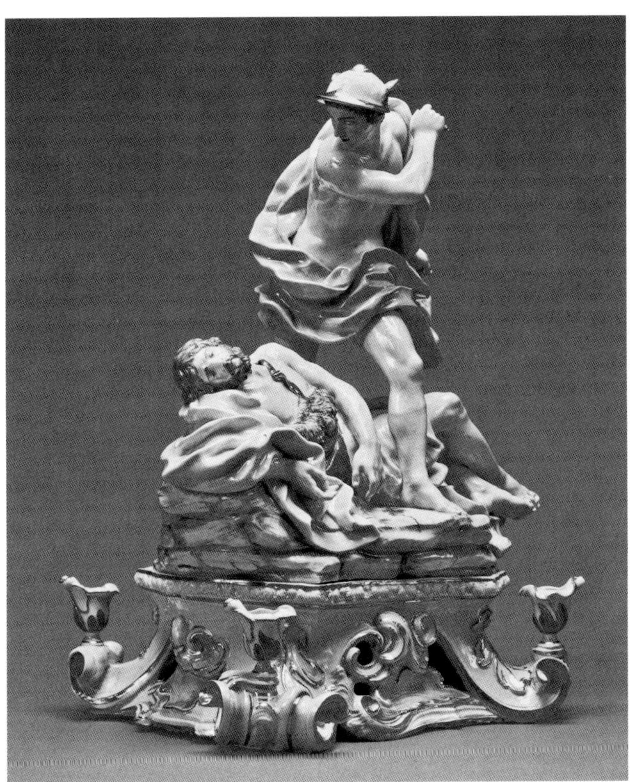

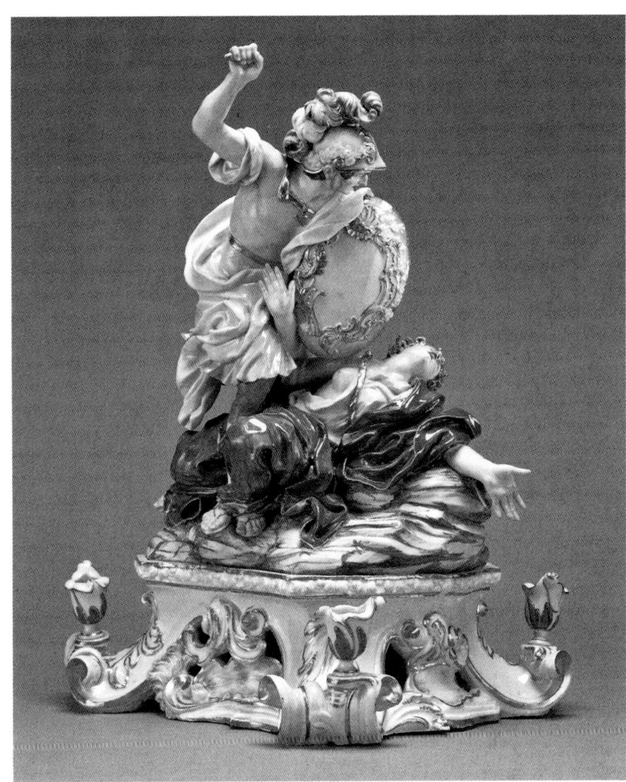

ABOUT THE ARTIST

The Ginori Porcelain Factory, one of the most important to be established in Italy, was founded in Doccia, near Florence, by the Marchese Carlo Ginori in 1735. Ginori hired two Viennese painters, as well as an Italian modeler, Gaspero Bruschi, to start production. For ten years, Ginori experimented with different porcelain recipes and collected models and molds for porcelain figures. Finally, in 1746, he began to sell the factory's products to the public. The early works were made from a grayish, hard-paste porcelain derived from local **clay**, which was extremely prone to cracking. After 1770 the paste was changed to a finer, whiter variety.

Under the directorship of Ginori's son, Lorenzo, the Ginori factory produced a range of polychromed and white bas-reliefs of mythological subjects. The single figures and figure groups included peasant, **classical,** and **commedia dell'arte** subjects. Ginori's was the only Italian porcelain factory to thrive during the 1800s; it continues production today under the name Richard-Ginori.

QUESTIONS FOR TEACHING

- What is happening in each sculpture?
- How are these two objects similar or different?
- What textures do you see? Where are they on the sculptures?
- What types of shapes are repeated in each sculpture?
- What natural motifs do you see?
- Have you seen sculptures like these before? Where did you see them? How were they used?

LESSON PLAN | **BEGINNING LEVEL**

Grades: Lower elementary (K–2), upper elementary (3–5)
Subjects: Science and visual arts
Time required: 3 class periods (plus a week to allow clay to dry)

Lesson Overview

Students study **porcelain** sculptures that have a functional purpose. They identify natural motifs and textures in these objects and then sculpt **clay** to create their own objects. They make observations about how a clay's **mixture** affects its properties.

Learning Objectives

Students will:

- Discover that clay is a mixture of water and very small rock particles.
- Observe the properties of clay and make conclusions about how a clay's mixture affects its properties.
- Identify textures, **shapes**, and natural motifs in works of art.
- Sculpt a decorative object out of clay that includes various textures, shapes, and natural motifs.

Materials

- Images of *Mercury and Argus* and *Perseus and Medusa* (p. 16)
- Information about the featured work of art and Questions for Teaching (pp. 88–89)
- Examples of porcelain **bisque-ware** and glazed porcelain
- Pencils
- Journal or bound paper
- Low-fire or high-fire clay, depending on availability of kiln (china clay/porcelain, if possible, or substitute oven-bake or self-hardening clay)
- Paper towels
- Magnifying glasses
- Student handout: Clay Observations (pp. 92–93)
- Cheesecloth or plastic bags
- A scale for weighing
- (Optional) Camera

Lesson Steps

1. Display images of the sculptures *Mercury and Argus* and *Perseus and Medusa*. Use the following questions to begin a discussion about the sculptures. Chart student responses.
   - What do you see?
   - What is the dominant color in the two objects?

2. Pass around various porcelain objects so that students can feel some of the qualities of porcelain, then continue discussing *Mercury and Argus* and *Perseus and Medusa*. Ask students the following:
   - If you could hold these sculptures, what would they feel like? What details in the artwork tell you this?
   - How would you describe the texture you see on the surface of the sculptures?
   - Based on your observations, what do you think these sculptures are made of?

   Review the chart to summarize what students observed about the artwork and what they think it is made of. Students will record observations and predictions in their journals.

3. Explain that the sculptures are made of china clay, a mixture of water and very small particles of **kaolinite** clay. Divide the class into groups of three. Hand out small samples of clay, paper towels, magnifying glasses, and the Clay Observations student handout. Students will use these materials to look closely at the clay and write about what they see. After students have finished their worksheets, have them share their findings. They should be able to identify the characteristics of china clay and describe how it is different from the two substances it contains: water and kaolinite clay.

4. Compare the observed clay characteristics to the students' observations from step 1 about the porcelain sculptures. Students should note differences, such as the difference between the hard sculptures and the soft and squishy clay. Explain that when china clay is heated by the sun, the water in the mixture evaporates and the clay dries out and becomes a solid. If the dry clay is then exposed to higher temperatures in a large oven known as a **kiln**, it goes through a chemical change to become porcelain. In a kiln, the **minerals** in the clay melt together to become very strong, hard, glasslike, and white.

5. Return to the images of the porcelain sculptures by the Ginori Porcelain Factory. Inform students that the sculptures were made at a time when many works of art were decorated with motifs inspired by nature. Tell students that

a motif is a shape, symbol, or design that is usually repeated in a work of art. Ask students what types of shapes are repeated in the sculptures. Then ask what natural motifs they see in the sculptures (e.g., leaves in the bases). Inform students that the bases of the sculptures have candle sockets and therefore the sculptures may have been part of a **candelabrum**.

6. Have the class use clay to sculpt functional decorative objects, such as candle holders, soap dishes, pencil cups, etc. Tell students to incorporate natural motifs and a variety of textures and shapes.

7. Lightly wrap each finished sculpture in cheesecloth or store it in an open plastic bag to let it dry slowly and evenly. Note that if smaller parts of the sculpture dry faster than the larger parts, they may crack or break off. Dry the sculptures in areas with different amounts of sunlight and heat: place some in a sunny window, some away from a window, and some in a dark area like a closet. Have the students check their sculptures every other day for a week to see how the clay changes as the water evaporates from the heat of the sun. Students should record their findings using the Clay Observations student handout. You may want to have students take photographs to record the changes in color.

8. After handouts are completed, discuss the class's findings and make conclusions about how the relative mixture of water and kaolinite in clay affects its properties. Students should be able to connect the changes in color, texture, and weight in the clay to the **evaporation** of water.

9. When the sculptures are dry, have students point out the natural motifs, textures, and shapes in each other's works. Also ask students to describe any changes they notice in the sculptures now that the clay has dried.

STUDENT HANDOUT

CLAY OBSERVATIONS, PAGE 1

Record your observations below as you examine a piece of wet clay.

1. Touch the clay sample. How does it feel? Describe the texture.

2. Using a magnifying glass, redescribe the texture.

3. Compare the appearance of the texture of the clay under the magnifying glass to the texture you see using only your eyes.

4. How is the clay similar to or different from water?

5. How is it similar to or different from soil?

6. What other observations do you have about the clay?

STUDENT HANDOUT

CLAY OBSERVATIONS, PAGE 2

Carefully observe the clay sculptures in each location (sun, shade, total darkness) on different days. Record your findings.

Location	Day 1	Day 3	Day 5
Sun	Color: Texture: Weight: Other:	Color: Texture: Weight: Other:	Color: Texture: Weight: Other:
Shade	Color: Texture: Weight: Other:	Color: Texture: Weight: Other:	Color: Texture: Weight: Other:
No Sun	Color: Texture: Weight: Other:	Color: Texture: Weight: Other:	Color: Texture: Weight: Other:

LESSON PLAN | **INTERMEDIATE LEVEL**

Grades: Middle school (6–8)
Subjects: Science and visual arts
Time required: 3–4 class periods (plus a week to allow clay to dry)

Lesson Overview

Students study **porcelain** sculptures that have a functional purpose. They identify natural motifs, **shapes**, and textures in these decorative art objects and then sculpt **clay** to create their own objects. Through the drying and firing of clay, they learn about the processes of **radiation**, **convection**, and **evaporation**.

Learning Objectives

Students will:

- Observe the properties of clay and make conclusions about how a clay's **mixture** affects its properties.
- Identify textures, shapes, and natural motifs in works of art.
- Sculpt a clay decorative art object that includes various textures and natural motifs.
- Understand the concepts of radiation, convection, and evaporation.

Materials

- Materials listed in the beginning-level lesson (p. 90)
- (Optional) **Kiln**

Lesson Steps

1. Complete the beginning-level lesson, adapting for grade level as appropriate.

2. While the students' sculptures are drying, discuss the two different processes that allow water to evaporate from the clay and then transform the clay into porcelain: radiation and convection.

3. The first process, radiation, occurs when the water in the mixture becomes hot from the **energy** of the sun. As the water is exposed to **heat** from the surface, its **atoms** and **molecules** begin to move quickly. This motion creates energy, which changes the water molecules into gas. Then the gas moves slowly to the surface of the clay and evaporates into the atmosphere. This makes the drying of clay a slow process.

   Tell students that dry, unfired clay is known as greenware and is very fragile. It is possible to turn greenware back into a mixture of wet clay by soaking it in water. Explain that the heat from the sun is not strong enough to create the permanent chemical change in the clay **minerals** required to make their clay sculptures hard and waterproof porcelain, like the porcelain of the *Mercury* and *Perseus* sculptures. Brainstorm with the class to come up with ideas for keeping the clay from becoming soft again, making it stronger so it won't break as easily, and making it waterproof. Chart responses.

4. The second process, convection, begins the transformation of clay into porcelain, and occurs when clay is baked in a kiln. As the clay is heated from all sides, the heat begins to cause motion and energy in the minerals of the clay in the same way that sunlight causes energy in water molecules during radiation. However, at first, the center of the clay remains cool, or still, because the energy has not yet moved to the center. Gradually, the surface heat moves into the clay, layer by layer. As it moves inward, it slows down as it interacts with cooler molecules. If the heat is applied slowly, the temperature at the center will gradually become the same as the surface temperature. If heated too quickly, the exterior molecules move faster than the interior molecules, and the surface may burn or crack.

   In order to be transformed into porcelain, clay must be heated to a temperature high enough to cause a chemical change in its minerals: 2192°F–2552°F. (That is seven times hotter than your oven gets when you bake cookies.) To maintain an unblemished surface and keep the clay from cracking, the heat must be raised slowly over many hours in increments of 300°F–400°F. It can take a few days to complete the process, called **bisque firing**. Once the clay is turned into bisque ware, it cannot be turned back into a wet clay mixture by adding water.

5. Have the students' dry sculptures fired in the school kiln. Students should note the time it takes to fire the clay and how often the temperature is raised. If you don't have access to a kiln, get examples from an art store of baked porcelain objects that have not been glaze fired. Compare the surface of the baked clay to that of the sun-dried clay. Chart student responses.

LESSON PLAN | **ADVANCED LEVEL**

Grades: High school (9–12)
Subjects: Science and visual arts
Time required: 4–5 class periods (plus a week to allow clay to dry)

Lesson Overview

Students study **porcelain** sculptures that have a functional purpose. They identify natural motifs and describe how elements of art are used in these decorative art objects. They sculpt **clay** to create their own decorative objects. Through the drying and firing of clay and an analysis of the effects of glazing, they learn about the processes of **radiation, convection, vitrification, evaporation,** and **condensation**.

Learning Objectives

Students will:

- Observe the properties of clay and make conclusions about how a clay's **mixture** affects its properties.
- Identify elements of art (including texture and **shape**) and natural motifs in a work of art.
- Sculpt and glaze a decorative object out of clay that includes various textures and natural motifs.
- Understand the concepts of radiation, convection, vitrification, evaporation, and condensation.

Materials

- Materials listed in the beginning-level lesson (p. 90)
- (Optional) **Kiln**
- (Optional) Ceramic **glazes** and acrylic synthetic paintbrushes

Lesson Steps

1. Complete the beginning-level and intermediate-level lessons, adapting for grade level as appropriate.

2. Look at the images of *Mercury and Argus* and *Perseus and Medusa* (p. 15). Ask the students the following questions: What other **colors** do you see besides white? How does the surface quality of these sculptures compare to the surface of the **bisque-fired** porcelain?

3. Explain that the porcelain was coated with a substance called glaze. Glazes are applied to ceramics to give the surface shine and protection and to add color. Glazes are mixtures of **minerals,** sometimes **pigments**, and **silica**. When these minerals are heated to high temperatures (1972°F–2269°F), they melt together with the silica to form a glasslike material. This process is called vitrification.

4. Explain the concepts of evaporation and condensation to students. Explain that **energy** is absorbed when a material evaporates or melts, whereas energy is released when a material condenses or freezes. During vitrification, heat and energy are absorbed into the minerals and silica in the glaze, causing the molecules to speed up and melt together, which spreads the pigment out over the porcelain. When it cools, the glaze loses energy and solidifies to become hard and adhered to the surface of the porcelain.

5. Have the students apply glazes to their bisque ware, then have their sculptures fired in the school kiln. Students should note the time it takes to glaze-fire the sculptures and how often the temperature is raised. If you don't have access to a kiln, provide examples of glazed porcelain objects. Compare the surface of the glazed-fired clay to that of the bisque-fired clay.

Fighting Corrosion to Save an Ancient Greek Bronze

Art historians and scientists work together to preserve works of art so that they may be enjoyed for years to come. Students study a bronze statue from antiquity that was found in the sea to understand how conservators remove and prevent corrosion caused by exposure to humidity.

Beginning Level
Students study an ancient bronze statue, analyze its pose, and learn about techniques used to conserve it. They employ observation and research to speculate about how the sculpture was lost at sea.

Intermediate Level
Students study an ancient bronze statue, analyze its pose, and learn about techniques used to conserve it. They learn that the bronze used to make this sculpture is an alloy of copper and tin with small amounts of antimony, lead, iron, silver, nickel, and cobalt. They use the periodic table to research the chemical formulas of compounds in bronze. Students compare conservation techniques for two ancient bronze objects.

Advanced Level
After completing sections of the beginning- and intermediate-level lessons, students learn about the oxidation-reduction reactions that occurred in the statue. Students speculate about the conservation techniques needed to preserve the bronze sculpture.

Standards Addressed
Refer to the charts for national and California state standards on the Getty website, www.getty.edu/education/teachers/classroom_resources/curricula/art_science/downloads/standards.pdf

FEATURED WORK OF ART ►

Victorious Youth
Greek, 300–100 B.C.
Bronze
59⅝ x 27 9/16 x 11 in.
J. Paul Getty Museum
77.AB.30
www.getty.edu/art/gettyguide/artObjectDetails?artobj=891

This statue depicts a naked youth standing with his weight on his right leg, crowning himself with an olive wreath. The olive wreath was the prize for a victor in the Olympic Games and identifies this youth as a victorious athlete. The eyes of the figure were originally inlaid with colored stone or **glass paste,** and the nipples and eyelashes were inlaid with **copper**, creating naturalistic color contrasts.

Found in the sea off the coast of Italy, this statue is one of the few life-size Greek **bronzes** that survive today, and it provides much information on the technology of ancient bronze casting. The place of origin of the statue is unknown, but either Olympia or the youth's hometown is possible. Romans probably carried the statue off from its original location during the first century B.C. or A.D., when Roman collecting of Greek art was at its height. The Roman ship carrying it to Italy may have sunk off the coast, preserving the statue for centuries in the sea.

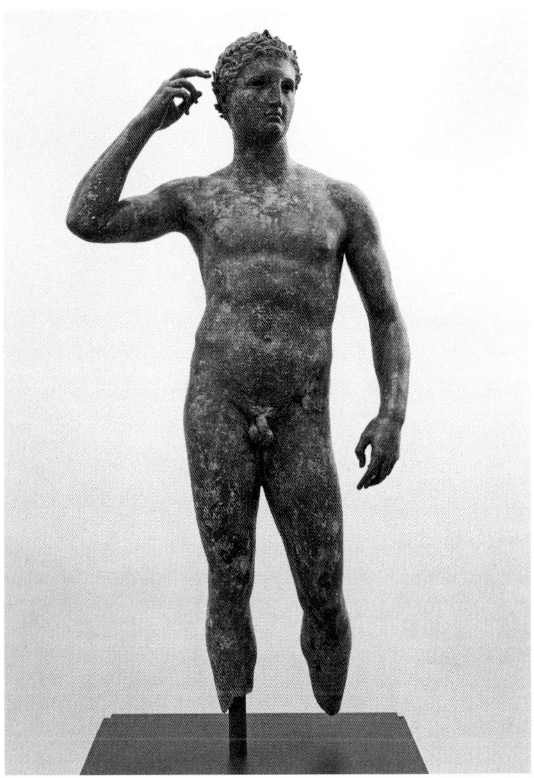

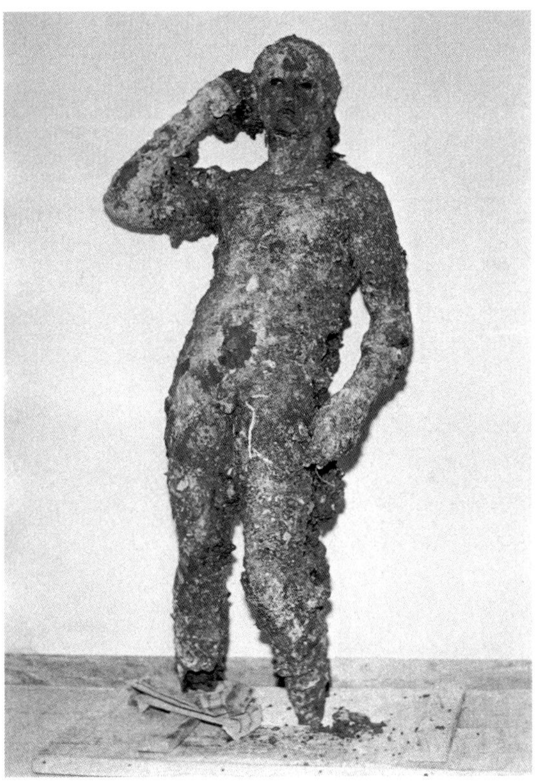

A thick layer of corrosion and incrustation covered the statue *Victorious Youth* when it was raised from the sea, as seen in this pre-conservation photograph.

QUESTIONS FOR TEACHING

- What do you notice about the pose of the figure?
- What is this person doing?
- Take a pose that you think represents the idea of victory. How is it similar to or different from this pose?
- What other details do you notice?
- What could be some reasons for the unusual colors on the surface of this sculpture?
- Look at a penny. What colors do you see? If you see green, what are some reasons the penny might have changed color? This sculpture is made of metals similar to those in a penny and reacts to water the same way a penny does if you toss and leave it in a fountain.

LESSON PLAN | **BEGINNING LEVEL**

Grades: Upper elementary (3–5)
Subjects: Science and visual arts
Time required: 2 class periods (plus one week for submerging metal in water)

Lesson Overview

Students study an object from **antiquity** that was found in the sea off the coast of Italy in order to understand how **conservators** remove and prevent corrosion on **bronze** statues. They derive meaning from analyzing the pose of the statue. Based on what they observe in the sculpture and what they read about the statue, students speculate about how the sculpture was lost at sea.

Learning Objectives

Students will:

- Observe and understand the changes that occur to **metals** when submerged in water.
- Analyze the pose of an ancient Greek sculpture.

Materials

- Image of *Victorious Youth* (p. 17)
- Copies for students of information about the featured work of art (pp. 96–97)
- Image of *Victorious Youth* before conservation (p. 18)
- Pennies
- Periodic table
- Journal or bound paper
- Clear plastic containers
- Water
- Teaspoons
- Salt

Lesson Steps

1. Divide the class into groups of three or four. Hand out a penny to each group. Explain that the metal in a penny is an **alloy** of zinc and copper. Have the class identify these two metals on a periodic table. Have students draw their pennies and then answer the following questions in their journals, using adjectives to describe characteristics.
   - What color is the coin?
   - Is it shiny or dull?
   - How would you describe the texture of the surface?
   - Is the metal hard or soft?

2. Have students place the pennies in clear plastic containers, cover them with water, and then add a few teaspoons of salt to the water. Allow the coins to soak for a week. Then have groups reexamine their coins. They should draw the pennies a second time and answer the questions above again in their journals. Each group should compare their two drawings and their observations. What has changed? What has stayed the same?

3. Display the pre-conservation view of *Victorious Youth*. Ask groups to discuss what they see in the image and to make guesses about what they may be looking at. Then explain to students that they are looking at an ancient Greek statue that was in the ocean for over two thousand years. Tell students that the statue is made of bronze. Explain to the class that over time the metal in their pennies would start to look like that of this statue, because the metal in the pennies would have a similar reaction to salt water. Explain that the statue has been affected by "bronze disease," which occurs when **chlorides** and **oxygen** combine with metal in a damp environment. The disease takes the form of a sudden outbreak of small patches of corrosion and is distinguished by rough, light-green spots. In this case, the damp environment was the ocean, and the chlorides and oxygen came from the salt water. **Mineral** deposits from the water and incrustations left by corals and other sea animals caused the rough layer on the surface.

4. Display the post-conservation view of *Victorious Youth*. Explain that scientists and conservators were able to stop the corrosion and remove the incrustations by cleaning the object. They used various tools and treatments. Over a three-month period, conservators scraped off the incrustations using mechanical tools to reveal the original surface. They also used several other methods to clean the object, such as extensive washing of the bronze with **chemicals** that neutralized the corrosion. Finally, they placed the statue in a relatively low-humidity environment. As a result, we are now able to get a better sense of what the object looked like originally.

5. Give students time to look at the sculpture and then ask them for their initial observations. Lead a discussion about the work using the following questions:
   - What can you tell about the sculpture now that the incrustations have been removed?
   - What do you notice about the pose of the figure?
   - What is this person doing?
   - What other details do you notice?

6. Share background information about the work of art. Based on the information provided, ask students to make up a story about how the sculpture might have been lost at sea.

LESSON PLAN | **INTERMEDIATE LEVEL**

Grades: Middle school (6–8)
Subjects: Science and visual arts
Time required: 3 class periods (plus one week for submerging metal in water)

Lesson Overview

Students study an ancient **bronze** statue, analyze its pose, and discover how **conservators** remove and prevent corrosion. They learn that the bronze used to make this sculpture is an **alloy** of **copper** and **tin** with small amounts of other **elements**. They use the periodic table to research the chemical formulas of **compounds** used to make bronze. Students compare conservation techniques in two ancient bronze objects.

Learning Objectives

Students will:

- Observe and understand the changes that occur to **metals** when they are submerged in water.
- Analyze the pose of an ancient Greek sculpture.
- Use a periodic table to identify the elements that compose bronze and understand the process by which these elements combine to form bronze.

Materials

- Materials listed in the beginning-level lesson (p. 98)
- Video: *Conserving Bronze: The Lamp with Erotes from Vani*, www.getty.edu/art/gettyguide/videoDetails?segid=4386

Lesson Steps

Note: This activity assumes that students have some prior knowledge of the periodic table and have had some exposure to basic **chemistry**.

1. Complete steps 1–5 of the beginning-level lesson and share information about the featured work of art.

2. Explain to students that bronze is a substance usually made of the elements copper and tin. The statue *Victorious Youth* was made of bronze containing copper, tin, and small amounts of antimony, lead, iron, silver, nickel, and cobalt. It also had patinas on its surface that contained copper and tin. Have students identify the elements copper and tin on the periodic table. Ask students to identify the characteristics of these elements based on their placement in the periodic table. Instruct students to research and record the electrical- and thermal-conductive properties of the two metals. Explain that the warm temperature of the seawater in which the statue was found sped up the corrosion of the metal. Ask students to identify which of the metals in the statue would be most affected by heat based on its thermal-conductive properties. Tell students that the combination of the corrosion layer and incrustations of sea life might have formed a protective exterior cocoon that preserved the interior and allowed the statue to last for over two thousand years.

3. Explain to the class that the tin and copper in the statue were compounds or alloys that were combined to make another compound, bronze. The tin was really tin **oxide** (SnO_2), and the copper was really cuprous oxide (CuO_2). Do not share the chemical formulas with the class. Instead, students should use a periodic table to identify the various elements and write the chemical formula for each compound. Explain that the two metals were melted and mixed together to form bronze. Bronze is created when high temperatures release the **oxygen**. Explain that bronze has its own properties that are distinct from those of copper and tin. Bronze oxidizes very slowly when exposed to air. It is less **malleable** than copper but has the same **conductivity**. Overall it has little **ductility**.

4. Have students view the video *Conserving Bronze: The Lamp with Erotes from Vani* on the Getty website. Ask students to describe the similarities and differences between the conservation techniques used to clean and restore *Victorious Youth* and the techniques used for the lamp in the video.

LESSON PLAN | **ADVANCED LEVEL**

Grades: High school (9–12)
Subjects: Science and visual arts
Time required: 1–3 class periods

Lesson Overview

Students study an ancient **bronze** statue, analyze its pose, and discover how **conservators** remove and prevent corrosion. They learn that the bronze used to make this sculpture is an **alloy** of **copper** and **tin** with small amounts of antimony, lead, iron, silver, nickel, and cobalt. They use the periodic table to research the chemical formulas of **compounds** used to make bronze. After learning about **oxidation-reduction reactions** that occurred in the statue, students speculate about the conservation techniques needed to conserve the bronze sculpture.

Learning Objectives

Students will:

- Observe and understand the changes that occur to **metals** when submerged in water.
- Analyze the pose of an ancient Greek sculpture.
- Use a periodic table to identify the **elements** that compose bronze and understand the process by which these elements combine to form bronze.
- Understand the oxidation-reduction reactions that occurred in a statue when it was made and then when it was exposed to seawater.

Materials

- Materials listed in the beginning-level lesson (p. 98)
- Video: *Conserving Bronze: The Lamp with Erotes from Vani,* www.getty.edu/art/gettyguide/videoDetails?segid=4386
- Digital projector with Internet access or a computer lab

Lesson Steps

1. Complete steps 1–5 of the beginning-level lesson and steps 2–3 of the intermediate-level lesson.

2. Explain to the class that some of the corrosion of *Victorious Youth* was caused by a **chemical reaction** between the bronze and the **oxygen** in the salt water. This chemical reaction resulted in an electron transfer known as **oxidation**. This means that the metal compound of bronze was combined chemically with the oxygen in the water to create an **oxide**. In this process, electrons were released at the **anode** (salt water) and taken up at the **cathode** (statue).

3. Have the class balance the oxidation-reduction reactions that occurred in the bronze when the statue was submerged in the ocean.
   • Electrochemical corrosion of copper alloy leads to the production of cuprous ions. Cuprous ions combine with the chloride in the seawater to form cuprous chloride, a major component of the corrosion layer:
   $Cu - e \rightarrow$
   Answer: $Cu - e \rightarrow Cu^+$
   $Cu^+ + Cl^- \rightarrow$
   Answer: $Cu^+ + Cl^- \rightarrow CuCl$
   • The statue now contains cuprous chlorides and is recovered and exposed to air. It continues to corrode when cuprous chlorides are combined with moisture and oxygen to hydrolyze and form hydrochloric acid and basic cupric chloride:
   $4CuCl + 4H_2O + O_2 \rightarrow$
   Answer: $4CuCl + 4H_2O + O_2 \rightarrow CuCl_2 \bullet 3Cu(OH)_2 + 2\,HCl$
   • The hydrochloric acid attacks the uncorroded metal to form more cuprous chloride:
   $2Cu + 2\,HCl \rightarrow$
   Answer: $2Cu + 2\,HCl \rightarrow 2CuCl + H_2$
   This process would have continued until no metal remained. Conservators had to scrape off the incrustation layers and find a way to stop the corrosion process.

4. Divide the class into working groups of four students. Explain that to stop the statue from corroding further, the conservators could have chosen among the following methods: electrolytic, electrochemical, sodium-sesquicarbonate, and sodium-sesquicarbonate/vacuum. Students can explore these treatments in detail in "Methods of Conserving Archaeological Material from Underwater Sites" on the Conservation Research Laboratory website, nautarch.tamu.edu/CRL/conservationmanual/. You may wish to note that the methods mentioned above are used to enhance salt extraction in archaeological metals and that, if the salt extraction is not completed, the reaction outlined in number 3 above would continue to occur.

5. Have groups come up with a combination of methods to conserve the bronze statue. Students should record findings in their journals. Groups will share their findings

with the class. Explain that the Getty conservators decided to use the sodium-sesquicarbonate/vacuum method to preserve the artwork. This method guaranteed that all unstable elements would regain stability without any loss to the original bronze or to the artistic qualities of the statue's surface. In addition, the statue is now kept in a humidity- and temperature-controlled room to prevent further corrosion. Compare students' recommendations to the actual treatment.

6. Return to the image of the sculpture and ask students to discuss the artistic qualities of the statue's surface. Which qualities do they think the **curators** were hoping to retain, and why would this be important?

Periodic Pigments

Students analyze a seventeenth-century still-life painting. They assemble and identify elements used to make pigment and then prepare pigment by grinding and heating ingredients, following a traditional method of paint making. They document and explain the chemical changes that occur in elements when they are used to make paint. Students understand the role of a binder in paint, experiment with two types of binders, and mix binder and pigment to make paint of desired quality. They use the paint they create in their own still-life paintings.

Beginning Level
Students make paint using raw umber and use it in their own monochromatic still-life paintings. They compare how long it takes for egg-based and oil-based paints to dry to better understand their properties.

Intermediate Level
Students make paint using raw umber and charcoal and use it to create two monochromatic still-life paintings. They hypothesize and test their hypotheses about the role of oxygen in the process of making charcoal.

Advanced Level
Students make paint using raw umber, charcoal, and cobalt blue. They compare the chemical changes that occur in various paint-making processes. They use the paint they create to make still-life paintings.

Standards Addressed
Refer to the charts for national and California state standards on the Getty website, www.getty.edu/education/teachers/classroom_resources/curricula/art_science/downloads/standards.pdf

FEATURED WORK OF ART ►

Ambrosius Bosschaert the Elder
Dutch, 1573–1621
Flower Still Life, 1614
Oil on copper
11¼ x 15 in.
J. Paul Getty Museum
83.PC.386
www.getty.edu/art/gettyguide/artObjectDetails?artobj=842

In this painting, a pink carnation, a white rose, and a yellow tulip with red stripes lie in front of a basket of brilliantly colored blossoms. Various types of flowers that would not bloom in the same season appear together here: rose, forget-me-not, lily-of-the-valley, cyclamen, violet, hyacinth, and tulip. Rendering meticulous detail, Ambrosius Bosschaert the Elder conveyed the silky texture of the petals, the prickliness of the rose thorns, and the fragility of opening buds. **Insects** crawl, alight, or perch on the bouquet. Each is carefully described and observed, from the dragonfly's transparent wings to the butterfly's minutely painted **antennae**. Although a vague reference, insects and short-lived flowers are a reminder of the brevity of life and the transience of its beauty.

A rising interest in **botany** and a passion for flowers led to an increase in painted floral still lifes at the end of the 1500s in both the Netherlands and Germany. Bosschaert was the first great Dutch specialist in fruit and flower painting and the head of a family of artists. He established a tradition that influenced an entire generation of fruit and flower painters in the Netherlands.

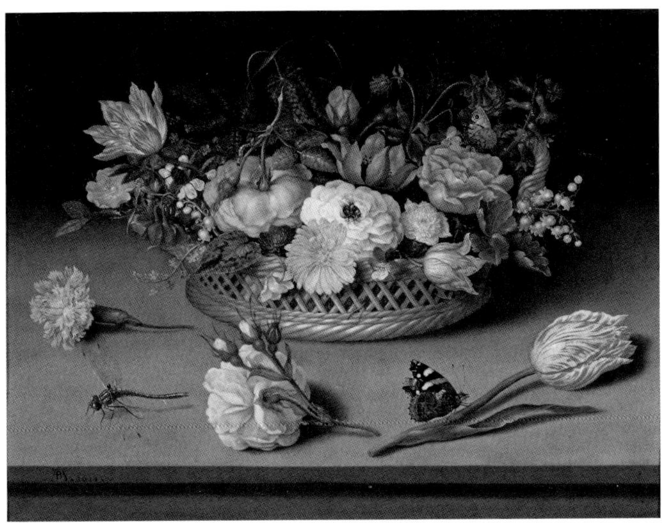

ABOUT THE ARTIST

During the 1600s the Dutch became Europe's leading **horticulturists**, and exotic flowers became a national obsession. Not surprisingly, flower painters were among the best-paid artists. In 1621, Ambrosius Bosschaert the Elder commanded a thousand **guilders** for a single flower picture. Nonetheless, his output of artworks was relatively small, for he was by trade an art dealer. Anticipating religious persecution, in 1587 Bosschaert's parents moved from Antwerp to Middelburg, a seaport and trading center second in importance only to Amsterdam. Six years later, Bosschaert joined Middelburg's **Guild of Saint Luke**. Bosschaert's works have been called flower portraits; each flower receives the same detailed attention as a face in a portrait. Usually small in scale and on **copper**, his paintings combined blooms from different seasons, painted from separate studies of each flower. It is not unusual to find the same flower, shell, or insect in many of his pictures. Like his predecessors, Bosschaert sometimes included symbolic or religious meanings in his works, such as the transience of life, by including objects at different stages in the life cycle.

QUESTIONS FOR TEACHING

- What objects do you see that are from nature? What objects do you see that are man-made?
- How many different kinds of flowers do you see?
- What are some reasons that an artist might have chosen to paint this subject?
- What colors did the artist use in this painting?
- Artists in Bosschaert's time were very interested in all aspects of the natural world. What evidence of this can you find in this painting?
- How does the inclusion of insects in the composition affect your interpretation of the painting?

PAINT AND PIGMENT

Paint is made by combining **pigment** (usually a finely ground powder) and **binder** (usually a liquid). Pigment gives paint its **color,** while binder holds the pigment and adheres it to a surface. Pigment particles are insoluble and are suspended in the binder. Pigments come from a variety of sources, including animals, vegetables, **minerals**, and synthetics. Some are earth pigments, or natural inorganic pigments, which are colored clumps of earth that have been ground and washed. These were the earliest pigments used and include **raw umber**, a brownish-yellow earth found in many places around the globe. Mineral pigments are another type of natural inorganic pigment and include colors such as **vermilion** (cinnabar/mercuric sulfide) and ultramarine (**lapis lazuli**).

Natural organic pigments are made of vegetable or animal products, rather than earth or minerals. These include such pigments as the now-outlawed Indian yellow, which was made in India from the dried urine of cows that were force-fed mango leaves. Other examples include **madder**, made from roots, and charcoal, made from carbonized wood. Artificial inorganic pigments, by contrast, are colors that are produced rather than found. Over the centuries, alchemists and then chemists invented many pigments, such as cobalt blue. These synthetics transformed the artist's palette by creating a new spectrum of colorful paints.

The binder is the adhesive material that gives body to paint and attaches it to a support such as canvas, parchment, panel, copper, paper, etc. The earliest binders were animal fat and wax. Later, egg yolk, egg white, honey, and tree sap were used as binders and resulted in water-soluble paints. Oils, particularly cold-pressed linseed oil, began to be used as a binder in Europe in the fifteenth century. Oil is slow to dry, is not water soluble, and produces a lustrous surface.

LESSON PLAN | **BEGINNING LEVEL**

Grades: Upper elementary (3–5)
Subjects: Science and visual arts
Time required: 2–3 class periods

Lesson Overview

Students make paint from **raw umber** and then use it to create monochromatic still-life paintings. They compare how long it takes for egg-based and oil-based paints to dry to better understand their properties.

Learning Objectives

Students will:

- Assemble and identify **elements** used to make **pigment**.
- Document and explain the chemical changes that occur in elements when they are made into paint.
- Understand the role of **binder** in paint and experiment with two types of paint binders.
- Use a mortar and pestle to grind pigment.
- Mix binder and pigment to make paint of desired quality.

Materials

- Reproduction of *Flower Still Life* by Ambrosius Bosschaert the Elder (p. 9)
- Information on the featured work of art and Questions for Teaching (pp. 102–3)
- Student handout: Watching Paint Dry (p. 107)
- 4 x 4 in. clayboard squares purchased at an art supply store or pieces of cardboard primed with gesso
- Pencils
- Mortar and pestle
- Raw umber (dry pigment)
- Synthetic paintbrushes, 1 inch or smaller
- (Optional) Magnifying glasses
- Egg **glair** (See the art activity "Making Paint" for instructions on how to make this, www.getty.edu/education/teachers/classroom_resources/tips_tools/downloads/making_paint.pdf)
- Vegetable oil (cold-pressed linseed is best)
- Paper or plastic cups or bowls
- Assorted objects from nature to use as subjects for students' still lifes
- Paper towels
- (Optional) Odorless paint thinner
- (Optional) Art activity: "Exploring Value to Create Form," www.getty.edu/education/teachers/classroom_resources/tips_tools/downloads/exploring_value.pdf

Lesson Steps

1. Display or hand out reproductions of *Flower Still Life* by Ambrosius Bosschaert the Elder. Discuss the work of art using Questions for Teaching as appropriate to grade level. Tell students that the paints used to create the original artwork were made by the artist in his studio, since paint was not available for sale in the seventeenth century. Each **color** was made using a different paint or mixture of paints. Ask students to identify all the colors they see in the painting. Create a chart of all colors mentioned.

2. Ask students to describe the characteristics of **liquids** and **solids** and discuss the differences between them. Using student input, write definitions for *liquid* and *solid* and list some examples of each on the board.

3. Explain what a **solution** is and ask for examples. Explain that paint is a solution because it is a mixture of a liquid and a solid.

Making Paint with Raw Umber

1. Give each pair of students a copy of the Watching Paint Dry student handout, a piece of clayboard (or cardboard primed with gesso), a pencil, a mortar containing one tablespoon of raw umber, a pestle, and a paintbrush. Ask students to examine the raw umber by looking at it, smelling it, and rubbing it between their fingers. You might also ask students to observe the material closely with magnifying glasses. Ask them where they think it comes from and discuss responses. Tell students that umber is a type of earth, and that many paints have colored earth as their bases. Explain that earth was the oldest material used to create paint and that it dates to prehistoric times. Ask students whether the umber is a liquid or a solid. Finally, tell students that the solid material used in paint is called pigment. Pigment gives paint its color. Many pigments are made from things that can be found in nature, like earth, **minerals**, rocks, plants, and even parts of animals and insects.

2. Ask students what liquid they think needs to be added to the umber to make paint. Explain that they will be adding two different liquids, or binders—oil and egg glair—to their pigment and documenting the differences in the resulting paints.

3. Have students draw a line across the center of their clayboards or cardboard and label one side "oil" and the other "egg." Have partners take turns grinding the pigment. Then they should put the ground pigment into a

bowl or cup. Add one-half teaspoon of oil to the umber in each pair's mortar. Have partners take turns mixing the pigment and oil together. It will seem dry at first but will slowly take on the consistency of mayonnaise. (This will take about ten minutes.) Once the oil paint is ready, have student pairs paint the entire "oil" side of the clayboard or cardboard with the oil-based paint using a large brush. Have students record the time and date on the Watching Paint Dry student handout.

4. Give each student pair a cup containing one teaspoon of umber and a clean paintbrush. Add a teaspoon of glair, made from egg whites, to the umber in each pair's cup. (View the art activity "Making Paint" on the Getty website for instructions on how to make glair.) Have students mix the materials until the umber has been incorporated into a smooth mixture. (This will take just a minute or two.) Students will paint the "egg" side of the clayboard or cardboard with this egg-based paint. After painting the board, students should note the time and date on the Watching Paint Dry student handout. Put the paint samples aside to dry, and clean up from the activity. Mortars and pestles should be cleaned with soap and water. Turpentine might be needed to wash oil-based paint out of the paintbrushes (this should be done by the teacher).

5. Have students monitor the paint samples and chart their drying times. The egg-based paint will dry very quickly; the oil-based paint will take a few days to dry.

6. Students should fill in the Watching Paint Dry student handout with their findings.

7. Discuss why the oil and egg glair dry at different speeds to address the concept of **evaporation**. Egg-based paint dries due to evaporation; oil-based paint dries through a process called **oxidation**. For more on the properties of oil paint, visit, "The Drying of Oils and Oil Paint" on the Smithsonian Museum Conservation Institute website www.si.edu/MCI/english/research/technical_studies/drying_oils_paint.html.

Painting a Still Life

1. Have students use the paint they made to create still-life paintings. Before they begin painting, they will warm up by sketching objects depicted in *Flower Still Life* by Ambrosius Bosschaert the Elder. Pass out copies of *Flower Still Life* or project an image of the painting in the classroom. Students will use pencils to sketch three objects from *Flower Still Life*. Students who are more advanced should pay attention to the negative space between the three objects and the negative space between the objects and the edges of the canvas.

2. Provide students with clayboard (or cardboard primed with gesso), paint, brushes, and paper towels. Have students select three objects from nature for their own still-life paintings. Tell more-advanced students to look closely at the objects and think about how they want to arrange the objects in their **compositions**. Have them consider the space between and around the objects when deciding on their arrangements. Once they decide on a composition, tell them to sketch an underdrawing (a preliminary outline of a composition that is used as a guide) on the clayboard or cardboard.

3. Tell students they will create a monochromatic painting (a painting using one color). They can apply more layers of paint to create darker shades. If they are using glair-based paint, they can add water to create lighter shades. (Odorless paint thinner can be added to the oil-based paint.) While less-advanced students can begin painting with their homemade paints, using their underdrawings to guide them, more-advanced students can refer to the painting by Bosschaert in order to focus on the use of **value**.

Exploring Value (for more-advanced students)

1. Point out the artist's use of value to create objects that look three-dimensional. Have students explore value as a warm-up before painting their still lifes. They will complete a value-scale exercise. Students will slowly build layers to create a gradual value shift across ten squares, moving from lightest to darkest as depicted below. Inform students that creating a value scale requires a bit of patience. The following instructions are for use with glair-based paint. (To use oil-based paint, substitute odorless paint thinner for water.)

- Load a synthetic brush with water and just a little paint. Roll the tip of the brush on a paper towel to remove excess paint.
- Square 1 should be left blank. Starting with square 2, use short parallel strokes to fill in the square with the lightest tint of color.
- Add a little more paint to each square. Square 10 should be as dark as possible. Example:

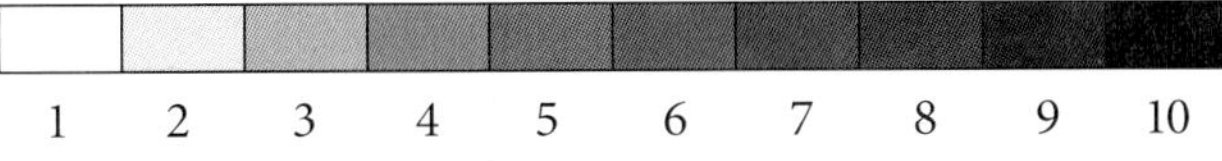

1 2 3 4 5 6 7 8 9 10

2. Now that students understand how to create gradations in value by using varying layers and mixtures of paint, they will create monochromatic compositions. Using their underdrawing to guide them, students should apply thin layers of paint to the surface, starting with the lightest shade. View the art activity "Exploring Value to Create Form" for ideas on how to use value to create the illusion of three-dimensional forms.

3. After paintings are completed, students should share them and discuss the challenges and rewards of making and using their own paint.

STUDENT HANDOUT

WATCHING PAINT DRY

Use this worksheet to record the times and dates you applied the two kinds of paint to the clayboard and to record when each of them dried. Compare and contrast the dry samples. Record your observations.

Paint	Time and Date of Application	Time and Date It Dried
Oil-Based Paint		
Egg-Based Paint		

1. Use the Venn diagram to record the similarities and differences between the two dry samples.

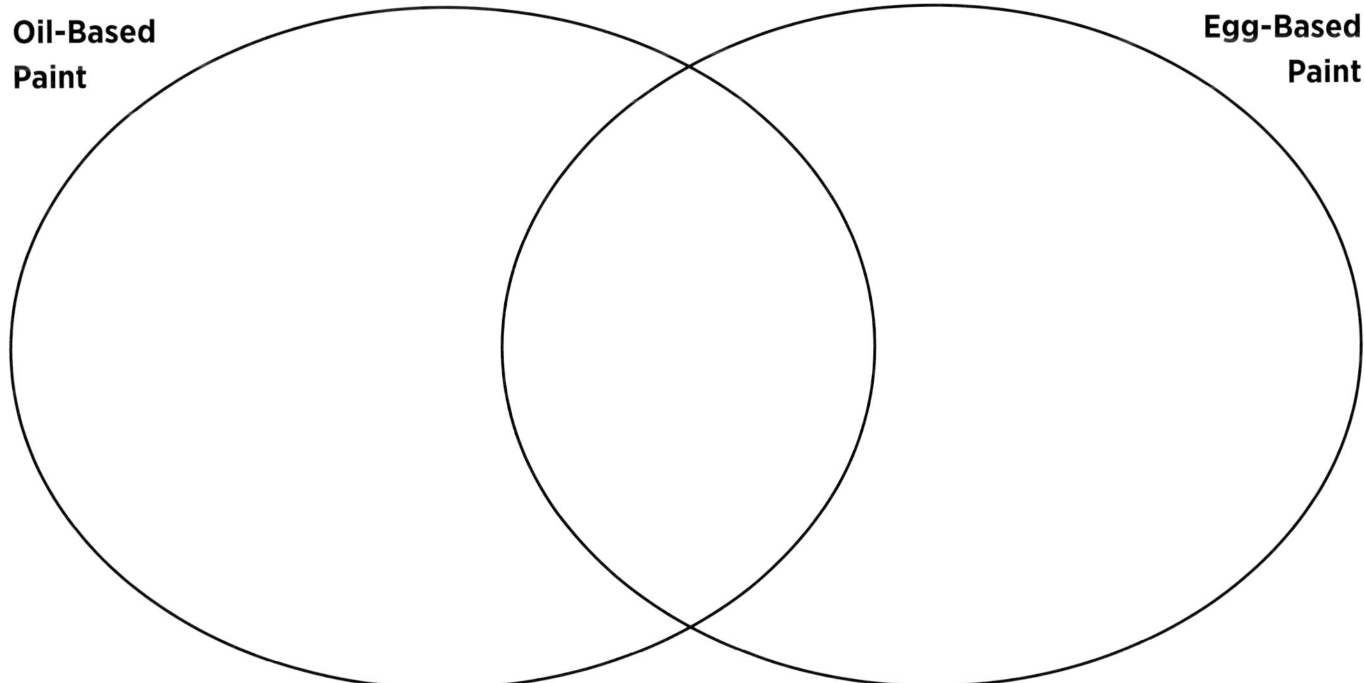

2. Speculate as to why they are similar:

3. Speculate as to why they are different:

LESSON PLAN | **INTERMEDIATE LEVEL**

Grades: Middle school (6–8)
Subjects: Science and visual arts
Time required: 3 class periods

Lesson Overview
Students make paint using **raw umber** and charcoal. They hypothesize and test their hypotheses about the role of **oxygen** in the process of making charcoal. Students use the paint they make in two monochromatic still-life paintings.

Learning Objectives
Students will:

- Assemble and identify **elements** used to make **pigment**.
- Prepare pigment by grinding and heating ingredients.
- Document and explain the chemical changes that occur in elements when they are used to make paint.
- Understand the role of **binders** in paint and experiment with two types of binders.
- Mix binder and pigment to make paint of desired quality.
- Hypothesize and test hypotheses about the role of oxygen in creating charcoal.
- Use a mortar and pestle.

Materials

- Materials listed in the beginning-level lesson (p. 104)
- Untreated pine dowels cut into pieces
- Crucibles
- Burners

Lesson Steps
Making Paint with Raw Umber

1. Complete steps 1–3 of the beginning-level lesson and steps 1–7 of the Making Paint with Raw Umber section of the beginning-level lesson.

Making Paint with Charcoal Black

1. Tell students they will be making charcoal, which they will then use as pigment in black paint. Explain that they will heat wood in two crucibles. Have students predict what will happen if one of the crucibles is left partially covered, rather than completely closed, while the wood is being heated. Students will test their hypotheses.

2. Place one or two pieces of pine in each of two crucibles. Mount the crucibles over burners. Cover one crucible tightly with a lid. Leave the other partially covered, with the lid ajar, to allow smoke to escape.

3. After two minutes, light gases will escape from around the lids of the crucibles. Once all gases have burned off, the charcoal is ready. Turn off the burners and let the crucibles sit until cool. Do not remove lids until cool.

4. Uncover crucibles and have students describe what they see, comparing the two products. The crucible with the closed lid will have charcoal inside, while the crucible that had the lid ajar will have ash. Have students speculate on the reason for the different results.

5. Remove the charcoal and grind it with a mortar and pestle until it is a fine powder. Sift the powder to remove large chunks and then place it in a cup or bowl. Add oil in a 1:4 ratio of charcoal to oil. Grind, or **mull,** the mixture until the paint has reached the consistency of thick cream or heavy soup.

6. Have students research the chemical changes that happen when charcoal is created.

7. Have a discussion in class, based on the findings from the research, about why the lack of oxygen in the crucible led to the transformation of wood into charcoal, and what gases were burned off during transformation.

Painting a Still Life

1. Complete the steps of the Painting a Still Life and Exploring Value sections of the beginning-level lesson, adapting for grade level as appropriate. For example, you might point out Bosschaert's use of overlapping shapes to create the illusion of space and encourage students to sketch overlapping objects that are arranged on a table.

2. Students will create another monochromatic **composition** with charcoal black. Students should use the layering technique, applying thin layers of paint to the surface, starting with the lightest shade. They should also consider leaving areas of the clayboard blank so that the color of the clayboard itself can act as the lightest value of the composition.

3. After paintings are completed, students should share them and discuss the challenges and rewards of making and using their own paint.

LESSON PLAN | **ADVANCED LEVEL**

Grades: High school (9–12)
Subjects: Science and visual arts
Time required: 3–4 class periods

Lesson Overview
Students make paint using **raw umber**, charcoal, and cobalt blue. They compare the chemical changes that occur when grinding **compounds** versus heating compounds. Students use the paint to create two monochromatic still-life paintings. IMPORTANT SAFETY NOTE: *Please be sure to implement this lesson in a properly ventilated room and have students wear appropriate masks and gloves.*

Learning Objectives
Students will:

- Assemble and identify **elements** used to make **pigment**.
- Prepare pigment by grinding and heating ingredients.
- Document and explain the chemical changes that occur in elements when they are used to make paint.
- Understand the role of **binders** in paint and experiment with two types of paint binders.
- Mix binder and pigment to make paint of desired quality.
- Hypothesize and test hypotheses about the role of **oxygen** in creating charcoal.
- Use a mortar and pestle.

Materials

- Materials listed in the beginning-level and intermediate-level lessons (pp. 104, 107)
- Masks and gloves
- Cobalt (II) chloride ($CoCl_2{\bullet}6H_20$)
- Aluminum chloride ($AlCl_3$)
- Test tube
- Burner

Lesson Steps

1. Complete steps 1–3 of the beginning-level lesson and steps 1–7 of the Making Paint with Raw Umber section of the beginning-level lesson.

2. Complete steps 1–7 of the Making Paint with Charcoal Black section of the intermediate-level lesson. Adapt for grade level as appropriate.

Making Paint with Cobalt Blue

1. Instruct students to wear masks and gloves.

2. Have student pairs measure out and combine the following in a mortar: 1 g cobalt (II) chloride ($CoCl_2{\bullet}6H_20$) and 5 g aluminum chloride ($AlCl_3$). Students should grind these together and place the mixture in a test tube.

3. Heat the mixture in the test tube over a gas burner until the reaction is completed and the mixture has turned blue. This should take about three to four minutes. Empty the contents into a mortar and let cool.

4. Grind the contents of the mortar and then transfer them to a bowl or cup. Add a few drops of linseed oil and mix until the paint is the consistency of thick cream or heavy soup. Add additional drops of linseed oil as needed to produce the proper consistency.

5. Students should document the process of creating the cobalt blue paint, research the chemical reaction that occurred to create the cobalt pigment, and answer the following questions:
   - What does grinding do? (physical change—breaks bonds)
   - What does heating do? (chemical change—re-bonds)

6. Have students use the three pigments they made to paint illustrations of the molecules used to make cobalt blue ($CoCl_2{\bullet}6H_20$ and $AlCl_3$).

Painting a Still Life

1. Complete the steps of the Painting a Still Life and Exploring Value sections of the beginning-level lesson, adapting for grade level as appropriate. For example, you might point out Bosschaert's use of overlapping shapes to create the illusion of space and encourage students to sketch overlapping objects that are arranged on a table.

2. After students create one monochromatic **composition** with the raw umber, they will create another composition with charcoal black and cobalt blue. (Note that the paint made with charcoal black and cobalt blue can be combined because they are both made with oil.) The blue and black paints can be combined to create a variety of shades of blue. Students should apply thin layers of paint to the surface, starting with the lightest shade. They should also consider leaving areas of the clayboard blank so that the color of the clayboard itself can act as the lightest value in the composition.

3. After paintings are completed, students should share them and discuss the challenges and rewards of making and using their own paint.

Objects of Mystery and the Scientific Method

Students practice scientific methodology to address the mystery of a carved Renaissance cabinet. They observe and describe the use of repetition, pattern, and balance in the design of a sixteenth-century cabinet and compare the design to that of a modern-day cabinet.

Beginning Level
Students follow the first step of the scientific method by making observations and writing descriptions of what they see. They sketch the sixteenth-century cabinet and a modern-day cabinet, incorporating repetition, pattern, and balance.

Intermediate Level
After comparing and contrasting cabinets from different time periods, students use observation skills and research to describe a sixteenth-century cabinet and speculate about what scientific processes could be used to determine its age.

Advanced Level
In addition to completing the activities in the intermediate-level lesson, students describe how carbon-14 dating can be used to date organic materials.

Standards Addressed
Refer to the charts for national and California state standards on the Getty website, www.getty.edu/education/teachers/classroom_resources/curricula/art_science/downloads/standards.pdf

FEATURED WORK OF ART ►

Cabinet
French, 1580
Carved walnut and oak with painted panels, linen and silk lining
121⅛ x 60⅜ x 22½ in.
J. Paul Getty Museum
71.DA.89
www.getty.edu/art/gettyguide/artObjectDetails?artobj=1104

Until 2001, most people believed this cabinet was made in the nineteenth century to look like a sixteenth-century cabinet. In other words, it was thought to be a fake. J. Paul Getty bought the cabinet in 1971 against the advice of his **curators**, who did not think it was genuine. The cabinet was in suspiciously pristine condition for one so old. The surface was coated with colored wax, suggesting that someone had tried to make it look older than it really was. Experts concluded that the cabinet was a product of the **Renaissance revival** of the nineteenth century, when American industrial magnates snapped up Renaissance-style furniture, including many fakes, from cash-strapped European aristocrats. Getty bought the cabinet for $1,700—much lower than its original asking price of $46,640 some fifty years before. The dubious cabinet was not displayed in the Getty Museum until 2006.

In 2001, curators in France organized an exhibition about the sixteenth-century cabinetmaker **Hugues Sambin**. After a Getty curator visited France to meet with scholars there, she decided to launch a thorough reexamination of the Museum's cabinet. Curators, **conservators**, and scientists teamed up to conduct research and scientific analysis. Science can help authenticate an object by dating the materials from which the object was made. When experts reexamined the cabinet in 2001, they came to a very different conclusion: that the cabinet was made in 1580 and is one of the rarest and most valuable cabinets from the French Renaissance in the United States.

The team's first scientific tool was **dendrochronology**, or tree ring dating. Using this method, they discovered that

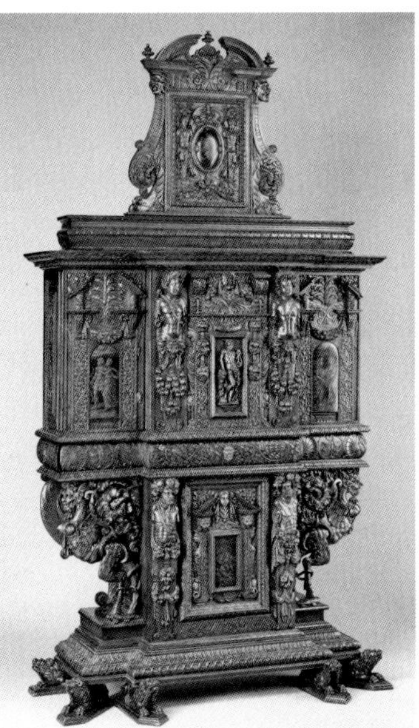

the oak tree used to make the cabinet's structural panels was harvested in late 1574 or early 1575 in Burgundy, a region in southeastern France. Factoring in the time needed to dry the wood and construct the cabinet, the team determined that the date of 1580 painted on the cabinet was indeed accurate.

Scientists then used **carbon dating** to date the surface wood of the cabinet. It is walnut, a wood with an irregular growth **pattern** that cannot be dated using dendrochronology. Scientists took tiny samples from the walnut and from the frayed edges of the silk and linen lining in the middle drawer. All the samples dated to between 1400 and 1600.

The team also used another tool: their eyes. They examined the marks left on the wood by woodworking tools. The rough marks on the back of the cabinet, for example, were made by a handheld saw and a type of woodworking tool known as a plane used in the sixteenth century. The walnut and oak parts of the cabinet also have identical marks from a **bench dog**, a toothed iron clip used to secure wood to a workbench. This suggests that both woods were worked on the same bench at roughly the same time. Under a layer of red velvet fabric in the cabinet, the team found old nail holes from an earlier lining. These holes held traces of the original sixteenth-century fabric that still lines the middle drawer. Analysis of the tacks and tack holes on the door of the cabinet provided further proof of authenticity.

The team delved into archival sources to learn more about the origins of the cabinet. Curators located a 1596 household inventory that describes two French Renaissance cabinets. One of the cabinets is now in the Museum of Time in Besançon, France, and has been definitively dated to 1581. The other cabinet may be the Getty cabinet. The paintings and carvings on the two cabinets were closely examined by curators. Minute clues and subtle coincidences of style indicated that the cabinets may have been made by the same artist or group of artists. Who was this person or persons? Curators, conservators, and scientists continue to work on this mystery.

Carbon-14 Dating

The **element** carbon exists naturally on Earth in two stable forms, or **isotopes**, known as carbon-12 and carbon-13. In carbon-13, the nucleus of the **atom** contains one extra subatomic particle called a neutron. A third isotope of carbon with two extra neutrons, known as carbon-14 (C-14), is produced in the atmosphere when atoms of nitrogen are blasted by high-energy cosmic rays streaming in from space. The C-14 atom is unstable—radioactive—and eventually decays back into nitrogen-14. Because some of the radioactive atom always exists at a low level in the atmosphere and on Earth, it gets incorporated at a predictable rate into the **cells** and tissues of all living things, from microscopic bacteria to plants to animals. For example, organisms ingest the C-14 containing carbon dioxide molecules as part of their natural metabolic processes.

Upon death, organisms cease taking in C-14, so as the isotope decays back to nitrogen-14, its presence steadily decreases over time. Measuring the amount of C-14 in something originally derived from organic material (such as the wood used to make the cabinet) tells the amount of time that has passed since the organism it came from (in this case, the walnut tree) was last alive.

What that means, however, is that any contamination from an object that was more recently alive will raise the relative abundance of C-14 atoms, implying a younger age. In the case of the Getty cabinet, scientists drilled into the wood and took samples from below the surface. This ensured that wax or polish made from organic materials applied in recent years would not contaminate the sample.

QUESTIONS FOR TEACHING

- This object is a piece of furniture. What do you think it might have been used for?
- What lines and shapes do you see?
- Which lines and shapes are repeated?
- Which elements form patterns?
- Which elements of the design on one side of the cabinet can you also find on the opposite side?
- What do you think it is made of?
- What sort of person might want or need this type of furniture?
- If this were a piece of furniture in your home, where might you place it?
- What components of this cabinet remind you of other cabinets you have see in someone's home or in magazines or movies? What components are different?

LESSON PLAN | **BEGINNING LEVEL**

Grades: Upper elementary (3–5)
Subjects: Science and visual arts
Time required: 1–3 class periods

Lesson Overview

Students describe how the scientific method was used to establish the age of a sixteenth-century cabinet. They observe and describe the use of repetition, **pattern**, and **balance** in the design of the cabinet and compare it to that of a modern-day cabinet.

Learning Objectives

Students will:

- Compare and contrast a sixteenth-century cabinet to a modern-day cabinet using observation and description skills.
- Understand that observation and description are part of the scientific method.
- Analyze how artists use repetition, pattern, and balance in decorative art objects.

Materials

- Image of the sixteenth-century cabinet (p. 19)
- Information on the featured work of art and Questions for Teaching (pp. 110–11)
- Images of contemporary cabinets (from magazines or catalogs)
- Paper and pencils
- Student handout: Comparing Cabinets (p. 114)

Lesson Steps

1. Tell students they are going to solve a mystery that puzzled **curators** and **conservators** at the J. Paul Getty Museum for more than fifty years. Display an image of the cabinet, or hand out reproductions, and tell students the following: Mr. Getty purchased this cabinet in 1971. At that time, nobody was sure when the cabinet had been made. Some people believed it was made in 1580, which meant it would have been more than three hundred years old. Others thought it was made much later.

   The Museum wanted to know the real age of the cabinet for many reasons. Knowing its correct age would help the Museum and its visitors understand its significance. Also, if the cabinet were really from 1580, it would be much more valuable, because cabinets from that time period are very, very rare.

   Art conservators, who are trained as scientists, follow the same scientific method of investigation that all scientists follow to learn about the world. Observation is the first step of the scientific method. So, the first thing the conservators did to solve this mystery was to look at the cabinet and compare it to other cabinets from different time periods, noting similarities and differences.

2. Lead a discussion about the design of the cabinet using the following questions:
   - What lines and shapes do you see?
   - Which lines and shapes are repeated?
   - Which elements form patterns?
   - Which elements of the design on one side of the cabinet can you find on the opposite side?

3. Ask the class to speculate about why Getty conservators would compare the cabinet to older cabinets. What would they learn if the cabinet looked a lot like other cabinets from 1580? What if the Getty cabinet looked nothing like any cabinets made in 1580? What would the conservators learn if the cabinet looked very similar to cabinets that were made in a different year, like 1880?

4. Divide the class into small groups and hand out images of modern cabinets. Have students discuss the designs of the modern cabinets using the questions from step 2.

5. Have students sketch the Getty cabinet and a modern cabinet. Students should fill their pages with their drawings. They should also incorporate repetition, pattern, and balance as appropriate to what they see in the cabinets' designs. Ask students what new discoveries they made about the cabinets after spending time closely looking at and drawing them.

6. Give the groups the Comparing Cabinets student handout. Tell students they will be following the first step of the scientific method by making observations and writing descriptions of what they see. Ask each group to compare the Getty cabinet with their modern cabinet and write down at least two similarities and two differences between them. Students can use their sketches to help with their comparisons.

7. Ask students to answer the following questions based on their comparisons:
   - Do you think the Getty cabinet was made at the same time as the other cabinet you looked at?
   - Based on your observations of the Getty cabinet, what would you need to do to the contemporary cabinet to make it look older?

   Discuss students' responses, encouraging them to back up their responses by stating what they *observe* and by using *descriptions,* rather than purely guessing. Point out to students that they have now used the first three steps of the scientific method: they have made observations, written descriptions of what they observed, and made hypotheses about the age of the cabinet based on their observations. The last step would be to test their hypotheses.

STUDENT HANDOUT

COMPARING CABINETS

Use the Venn diagram to compare and contrast the Getty cabinet to a modern cabinet.
Use your sketches to help with your observations.

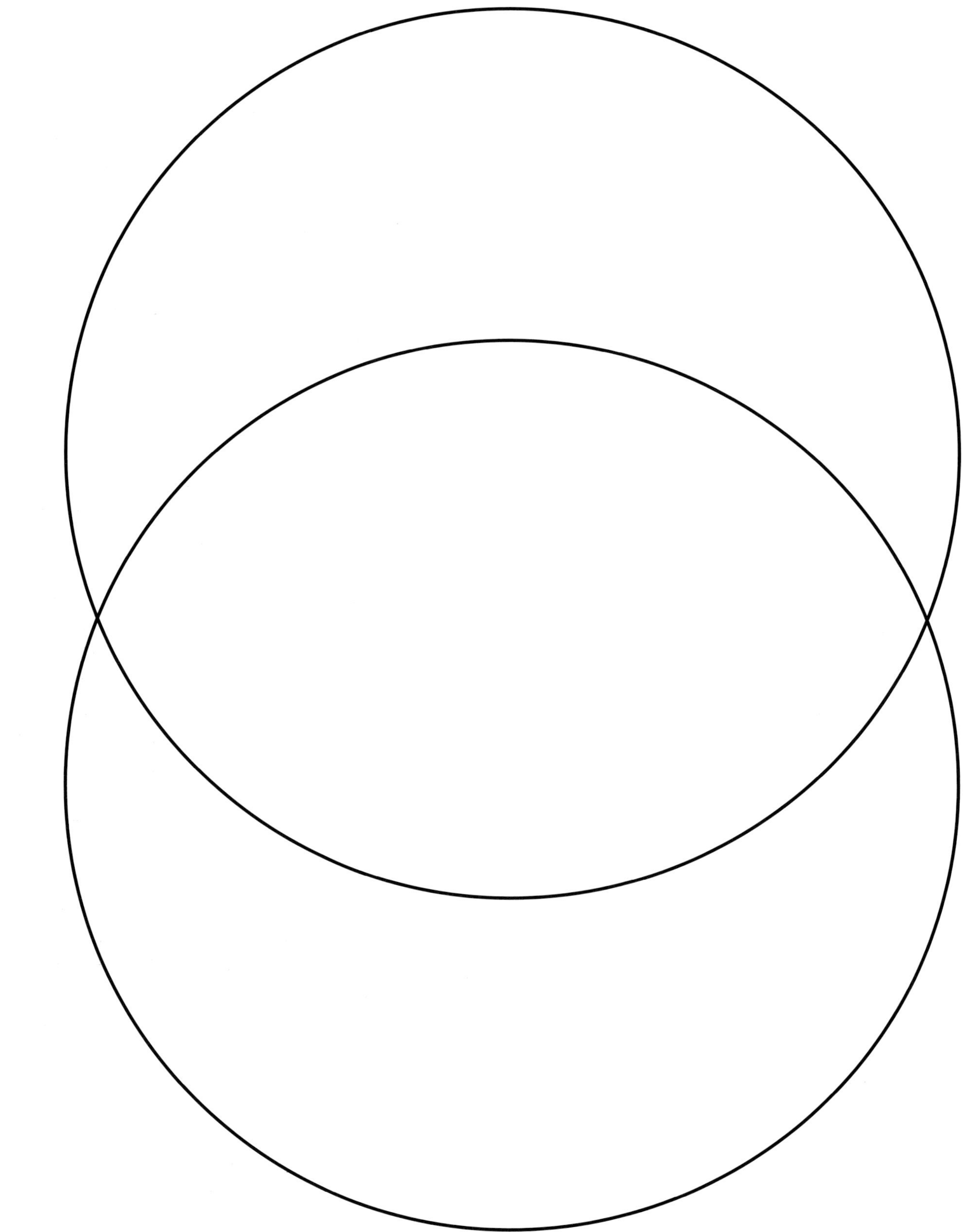

LESSON PLAN | **INTERMEDIATE LEVEL**

Grades: Middle school (6–8)
Subjects: Science and visual arts
Time required: 1–3 class periods

Lesson Overview

Students describe how the scientific method was used to establish the age of a sixteenth-century cabinet. They observe and describe the use of repetition, **pattern**, and **balance** in the design of the cabinet and compare it to that of a modern-day cabinet. Based on observation and information provided about the cabinet, students speculate about what scientific processes could be used to determine its age.

Learning Objectives

Students will:

- Analyze how artists use repetition, pattern, and balance in decorative art objects.
- Compare and contrast a sixteenth-century cabinet to a modern-day cabinet using observation and description skills.
- Speculate about how the age of the cabinet might be determined.
- Understand that the scientific method involves four steps: observation, description, developing hypotheses, and testing hypotheses.

Materials

- Materials listed in the beginning-level lesson (p. 112)
- Information from the exhibition *A Renaissance Cabinet Rediscovered,* www.getty.edu/art/exhibitions/cabinet/

Lesson Steps

1. Complete steps 1–5 of the beginning-level lesson, adapting for grade level as appropriate. For example, when viewing images of the cabinets, students might discuss other art and design elements in addition to repetition, pattern, and balance, such as **color**, materials, and functionality.

2. Return to the image of the Getty cabinet. Tell students that the cabinet was of unknown origin and that stylistic comparisons and research into its **provenance** were inconclusive. Explain to students that scholars had developed two theories, or hypotheses: one was that the cabinet was an authentic sixteenth-century object from Burgundy in France, and the other was that this cabinet was made in the late nineteenth century but was carved and **varnished** to make it look older (in other words, it was a fake).

3. Ask students to brainstorm about how science might solve this mystery and establish the age of the cabinet. Ask students to discuss with a partner what steps they might take to determine the age of the cabinet.

4. Have students view the online exhibition *A Renaissance Cabinet Rediscovered* on the Getty website to see how **conservators** and **curators** at the Getty determined the age of the cabinet using scientific methods.

5. Discuss the four steps of the scientific method with students, then ask them to identify the four steps taken by the Getty's conservators. What observations did conservators make about the cabinet? How did they describe the cabinet? What was the hypothesis the conservators were testing? Finally students should list the methods conservators used to test their hypothesis.

6. Explain to students that the scientific method is a standard process used by all scientists in order to ensure that research is done in a consistent and objective way. Lead a discussion about the philosophy of science by asking students if they think the scientific method can guarantee that scientific results will be reliable and objective. Have students make a list of questions that could be asked of scientific studies in order to assess the reliability of the data. Questions might include:
   - Who is doing the study? Is there any possibility of a bias?
   - Is this research repeatable?
   - Is the appropriate sample being studied?
   - Are there any controls in place?

LESSON PLAN | **ADVANCED LEVEL**

Grades: High school (9–12)
Subjects: Science and visual arts
Time required: 1–3 class periods

Lesson Overview

Students describe how the scientific method was used to establish the age of a sixteenth-century cabinet. They observe and describe the use of repetition, **pattern**, and **balance** in the design of the cabinet and compare it to that of a modern-day cabinet. Based on observation and information provided about the cabinet, students speculate about which scientific processes might be used to determine the age of the cabinet. They learn about the process of **carbon-14 dating** and how it can be used to date organic materials such as those in the cabinet.

Learning Objectives

Students will:

- Analyze how artists use repetition, pattern, and balance in a decorative art object.
- Compare and contrast a sixteenth-century cabinet to a modern-day cabinet using observation and description skills.
- Speculate about how the age of the cabinet might be determined.
- Understand that the scientific method involves four steps: observation, description, developing hypotheses, and testing hypotheses.
- Understand the **physics** concept of half-life.
- Describe how carbon-14 (C-14) dating can be used to date organic materials.

Materials

- Materials listed in the beginning-level lesson (p. 112)
- Information about carbon-14 dating (p. 111)
- Sets of one hundred M&M'S
- Small boxes with lids
- Student handout: Finding the Half-Life (pp. 117–18)

Lesson Steps

1. Complete steps 1–5 of the beginning-level lesson and steps 2–4 of the intermediate-level lesson, adapting for grade level as appropriate. For example, when viewing images of cabinets, students might discuss other art and design elements in addition to repetition, pattern, and balance, such as **color**, materials, and functionality.

2. Introduce students to the process of using radioactive carbon-14 to determine the age of organic materials. Share the information about carbon-14 dating. Continue with lecture and notes, discussion, or exploration of the "Carbon-14 Dating" section of the NDT (Nondestructive Testing) website, www.ndt-ed.org/EducationResources/CommunityCollege/Radiography/Physics/carbondating.htm.

3. As a class, discuss the following:
   - What is carbon-14?
   - How does it get incorporated into living things?
   - What does it mean when something is radioactive?
   - Does radioactive decay occur randomly or at a predictable rate?
   - How can carbon-14 dating be used to determine the age of organic matter?

   Students should understand that C-14 dating uses the known half-life of carbon-14 in organic materials to calculate the time of an organism's death.

4. Divide the class into groups and introduce the following activity as an analogy for the half-life principle, so that students will consider what they know about using the predictable nature of radioactive **isotopes** while performing the activity.

5. Provide students with sets of one hundred M&M'S, boxes with lids, and the Finding the Half-Life student handout.

6. Instruct students to complete the half-life activity described on the handout and create a class chart to record the results of each group's experiment. As students perform the activity, ask them to name how they are performing steps of the scientific method—observation, description, and developing and testing hypotheses. They can repeat the activity in order to test the validity of the results. Comparing the results of different student groups' experiments is another way of testing the data.

7. Lead a discussion that connects the half-life activity with radioactive isotopes. Have students describe how the radioactive isotopes have a predictable nature and that scientists can use this concept to help them find out the age of various objects, including the Getty cabinet.

8. Complete steps 5–6 of the intermediate-level lesson.

STUDENT HANDOUT

FINDING THE HALF-LIFE, PAGE 1

Directions for each group:

1. Place the M&M'S in the bottom of the box such that all candies are "M"-side down.

2. These candies exist in two "states": "M"-side up and "M"-side down. Your group will keep track of which is which. Use this table to record your observations.

Trial #	# of Candies "M"-side Down
0	100
1	
2	
3	
4	
5	
6	
7	
8	
9	
10	

3. Close the box and shake it *gently* up and down three times.

4. Open the box and remove all the candies that are now "M"-side up. Count and record on the chart above the number of candies that remain in the box ("M"-side down).

5. Repeat step 3 through step 5 until all the candies are out of the box.

STUDENT HANDOUT

FINDING THE HALF-LIFE, PAGE 2

6. Graph the information from your chart.

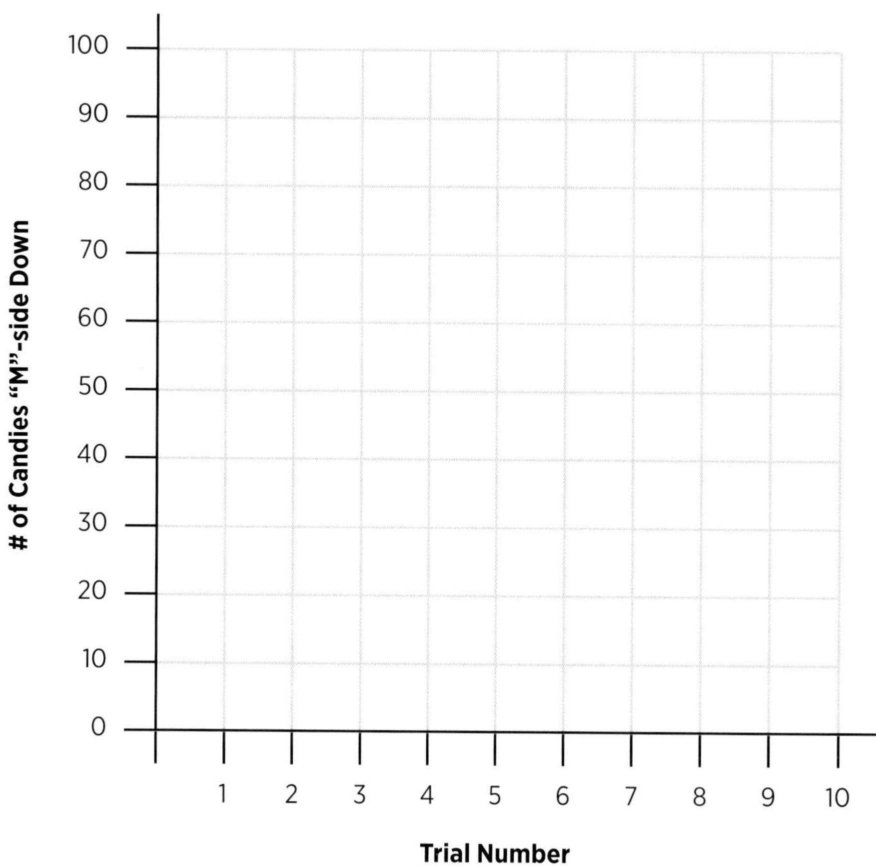

7. If each trial represented one hundred years, how long did it take for half of the candies (your carbon-14 atoms) to "decay"?

8. What would happen to the graph if you increased the number of candies (atoms) in the box?

Timeline

B.C.

- 500–480
The **direct lost-wax method** for **casting bronze** sculpture is developed. (Greece)

- 323
The mathematician Euclid writes *Elements*, a text on geometry and number theory. (Greece)

- 300–100
Victorious Youth (p. 17) is cast. (Greece)

- 250
Parchment is produced from the dried skins of animals such as calves, goats, and sheep. (Pergamum [present-day Turkey])

A.D.

- 105
Ts'ai Lun, an imperial court official, invents the paper-making process, which later spreads east and west along the Silk Road. (China)

- About 150
Ptolemy, the scientist, astronomer, mathematician, and geographer, writes the *Almagest*, the earliest surviving astronomical treatise. (Greece)

- About 400
The *Vergilius Vaticanus* (*Vatican Virgil*), an illuminated manuscript containing the writings of the poet Virgil (died 19 B.C.), is created. (Rome)

- 600
The printing of books begins with engraved woodblocks. (China)

- Between 600 and 850
The Mayans build El Caracol observatory at Chichen Itza. (Mexico)

- 782
The scientist Jābir ibn Hayyān introduces experimental investigation into **alchemy** and creates the basis of modern **chemistry** and the scientific method. (Kufa [present-day Iraq])

- Early 1200s
Astronomical Miscellany (p. 10) is produced and illustrated. (England)

- 1275
The physician William of Saliceto writes *Chirurgia* (*Surgery*), which documents his work with human dissection. (Italy)

- 1364
The Aztecs build Tenochtitlan. (Mexico)

- 1412
The architect Filippo Brunelleschi writes *Rules of Perspective*, a text describing the mathematical laws related to linear perspective. (Italy)

- 1450
Florence becomes the center of Renaissance culture. (Italy)

- 1460–70
The French King at Court (p. 15) from *The Story of Two Lovers* (*Historia de duobus amantibus*), an illuminated manuscript of the novel by Eneas Silvius Piccolomini (Pope Pius II), is created. (France)

- 1490
The artist and inventor Leonardo da Vinci draws *The Vitruvian Man*. (Italy)

- 1500
Faience and **majolica** are first produced industrially. (Italy)

- 1519
The explorer Ferdinand Magellan leaves on his voyage to find a route from Europe to Asia by sailing across the Atlantic Ocean. (Spain)

- 1537
The physician and alchemist Paracelsus writes *Grosse Astronomie* (*Whole Astronomy*), a manual of astrology. (Switzerland)

- 1542
Andreas Vesalius, a physician and the founder of the modern study of anatomy, writes *De fabrica corporis humani* (*On the Structure of the Human Body*). (The Netherlands)

- 1547
The astrologer Nostradamus makes his first predictions. (France)

- About 1550
Oval Basin (p. 8) is made by Bernard Palissy. (France)

- 1575
Craftsmen in Venice and Florence make the first attempts at imitating Chinese **porcelain**. (Italy)

- 1580
The French Renaissance Cabinet (p. 19) is carved in Burgundy. (France)

- 1590
Galileo Galilei, the mathematician and astronomer, writes *De motu* (*On Motion*), a description of experiments he conducted to determine the rate of speed of falling objects. (Italy)

- 1592
The ancient Roman city of Pompeii, buried by the volcanic eruption of Mount Vesuvius in A.D. 79, is rediscovered. (Italy)

- 1602

The astronomer Tycho Brahe's *Astronomiae instaratae progymnasmata* (*Introductory Exercises toward a Restored Astronomy*), which details the location of 777 fixed stars, is published (edited by Brahe's assistant, the astronomer Johannes Kepler). (Prague)

Galileo investigates the properties of pendulums. (Italy)

- 1608

The lensmaker Johann Lippershey invents the telescope. (The Netherlands)

Galileo constructs an astronomical telescope. (Italy)

- 1610

The alchemist Jean Beguin writes *Tyrocinium chymicum* (*Beginner's Chemistry*), the first chemistry textbook. (France)

- 1610–15

Juggling Man (p. 13) is created by Adriaen de Vries. (Prague)

- 1614

Flower Still Life (p. 9) is painted by Ambrosius Bosschaert the Elder. (The Netherlands)

- 1616

The astronomer and mathematician Willebrord Snell discovers Snell's law of **refraction**. (The Netherlands)

- 1620

Plymouth colony is established. (America)

- 1624

The chemist Johannes Baptista van Helmont identifies gases as a substance. (Belgium)

- 1637

The philosopher and mathematician Rene Descartes writes *La Géométrie* (*Geometry*). (France)

- 1650–55

Butterfly, Caterpillar, Moth, Insects, and Currants (p. 7) is drawn by Jan van Kessel. (The Netherlands)

- 1651

The astronomer Giovanni Riccioli names many lunar features in his map of the moon. (Italy)

- 1657

The scientist Christiann Huygens, following the research of Galileo, creates the design for the first clock pendulum. (The Netherlands)

- 1665

The scientist **Robert Hooke** coins the term ***cell***, to describe the **biological** structures in cork, in his ***Micrographia*** (also titled *Some Physiological Descriptions of Minute Bodies Made by Magnifying Glasses*). (England)

- 1683

Isaac Newton, the physicist and mathematician, explains his theory of the gravitational attraction on the tides by the sun, moon, and earth. (England)

- 1749

Mercury and Argus and *Perseus and Medusa* (p. 16) are produced by the Ginori Porcelain Factory. (Italy)

- About 1751

Compound Microscope and Case (p. 12) is made by Jacques Caffieri. (France)

- 1775

The American Revolution begins (ending in 1783 with British recognition of U.S. independence). (America)

- 1794

The inventor Eli Whitney patents the cotton gin. (America)

- 1807

The scientist **William Hyde Wollaston** invents the **camera lucida**. (England)

- 1816

Sir David Brewster, the mathematician and astronomer, invents the kaleidoscope. (Scotland)

- 1830

The botanist Robert Brown discovers the cell nucleus in plants. (Scotland)

- 1839

The artist and physicist **Louis-Jacques-Mandé Daguerre** announces the invention of the **daguerreotype** method of **photography**. (France)

- 1848

Gold is discovered in California. (America)

- 1857

The chemist and microbiologist Louis Pasteur proves that fermentation is caused by biological organisms. (France)

- 1857–60

The Emperor's Private Mosque in the Marble Palace, Agra Fort, India (p. 11) is photographed by Dr. John Murray. (India)

- 1859

The naturalist Charles Darwin publishes his theory of evolution in *On the Origin of Species by Means of Natural Selection*. (England)

To learn more about the broader historical context surrounding the works of art featured in this curriculum, see *The Timetables of History: A Horizontal Linkage of People and Events* by Bernard Grun, 4th revised edition (New York: Simon & Schuster, 2005).

Glossary

abdomen: The posterior part of the body of an insect, attached to the thorax.

abiotic: Characterized by the absence of living organisms.

alchemy: A speculative practice, elements of which led to the development of the fields of chemistry and medicine, that focused on the conversion of base metals into gold and the finding of a solution for everlasting youth.

allegorical: Related to allegory, a literary and artistic means of using fictional stories, characters, and events to reveal truths about human existence.

alloy: A mixture of two or more metals, combined to add strength or resist corrosion.

ambrotype: An early type of photograph in which an underexposed negative on glass was backed with a black coating, thus creating a positive image.

anode: In an oxidation-reduction reaction, the point where oxidation occurs (where a chemical compound loses electrons).

antennae: A pair of long, thin feelers on the heads of insects.

antiquity: Ancient times, especially those before the beginning of the Middle Ages (A.D. 476).

astronomy: The study of the characteristics and actions of objects and material in outer space.

atom: The smallest unit of a chemical element.

balance: The way a work of art's elements are arranged; also, the way weight is distributed so that a person or thing remains upright and steady.

Baroque: The principal European style in the visual arts in the seventeenth century and first half of the eighteenth century, between Mannerism and the Rococo, characterized by dynamic movement, monumental scale, and overt emotionalism.

bellows: A device that, by alternately expanding and contracting an air chamber, creates a stream of air through a nozzle.

bench dog: An iron clamp with teeth, used to secure wood to a workbench.

binder: A substance, such as egg or oil, that promotes the cohesion of pigments.

binomial nomenclature: The two-name system for classifying living organisms, the first one indicating the genus (scientific family); the second, the specific category (species).

biodiversity: The variety of living organisms in a particular environment.

biological: Related to biology, the study of living organisms and the processes of life.

biotic: Characterized by the presence of living organisms.

bisque firing: An initial firing of clay, before glazing, to cause vitrification into a ceramic state.

bodycolor: Watercolor mixed with opaque white pigment.

botany: The study of plant life.

bronze: An alloy of copper, tin, and sometimes other elements, often used to create sculptures; also, a sculpture made of this metal.

Brueghel the Elder, Jan: (Flemish, 1568–1625) Born in Brussels and trained by his grandmother, Brueghel was a draftsman and painter who worked from nature. He was celebrated in his own time for his skill at painting rich and delicate textures of flowers and landscapes. He frequently provided lush, warm-toned woodland scenes densely populated with exotic animals and flowers as frames for other artists' figures.

Brueghel the Younger, Jan: (Flemish, 1601–1678) Draftsman and painter from Antwerp, who devoted his career to carrying on the painting style of his father, Jan Brueghel the Elder. He sometimes copied his father's works and sold them under his father's signature, making it difficult to distinguish their styles, although Jan the Younger's few dated pictures show lighter colors and less precise drawing. He is best known for landscapes with villages, mythological scenes, allegories, and animals. His allegories depicted the senses, elements, seasons, and abundance.

cabinet de curiosité: In seventeenth-century Europe, a room for the research and display of collections of natural specimens, works of art, manuscripts, and other objects, usually in the homes of wealthy and/or aristocratic collectors.

camera: A device used for capturing visual images.

camera lucida: An optical instrument, invented by William Hyde Wollaston, that casts an image of an object onto a surface through the use of a prism or mirrors so that the object can be traced.

candelabrum: A candleholder with several arms.

carbon-14 dating (carbon dating): The process by which the age of ancient organic materials can be determined by measuring the rate of radioactive decay in their carbon-14 content.

carnivore: An animal that feeds on flesh.

cast: To form an object by pouring a liquid material into a mold and letting it harden; also, an object made by this method.

cathode: In an oxidation-reduction reaction, the point where reduction occurs (where a chemical compound gains electrons).

cell: An organism's smallest biological structure, usually microscopic, containing a nucleus surrounded by a membrane; often called the building block of life.

centrifugal force: The force acting on an object moving in a circle to pull it outward from the center of rotation.

centripetal force: The force acting on an object moving in a circle to keep it moving toward the center of rotation.

ceramic: Material made by shaping and firing clay.

chemical: A substance with a particular composition, structure, properties, and behavior.

chemical reaction: The interaction of two of more chemicals, which alters their properties by forming or destroying the bonds between atoms.

chemistry: The study of the composition, structure, properties, and behavior of substances.

chloride: A compound of chlorine and another chemical element; a common example is sodium chloride (salt).

classical: Related to the culture, art, literature, or ideals of the ancient Greek and Roman world.

clay: A stiff, often sticky, impermeable layer of soil that can be molded into shapes when wet; used for bricks, ceramics, and sculpture.

color: The hue, intensity, and value of the appearance of an object, as perceived by light receptors in the eye; one of the elements of art.

commedia dell'arte: Italian for "comedy of skill," popular in the sixteenth and seventeenth centuries, where actors playing stock characters improvised their roles in standardized plots.

composition: The arrangement of elements in a work of art.

compound: A chemical combination of two or more elements that cannot be redivided into separate components.

compound microscope: A microscope with two or more convex lenses, providing higher magnifications of a specimen.

condensation: The process by which a gas is converted to a liquid.

conductivity: The ability of a material to transmit electricity, heat, or sound.

conservator: A person responsible for the preservation of items of cultural, historical, or environmental value.

constellation: A formation of stars in a configuration perceived as one of eighty-eight particular designs, including Ursa Major.

consumer: In an ecosystem, an organism that consumes materials created by other organisms.

convection: The process by which heat is transferred throughout a liquid or gas through the movement of its molecules.

convex lens: A lens that curves outward and refracts light toward the center of the lens, providing focus.

copper: A ductile, malleable, reddish-brown metallic element used both by itself and combined with tin to make bronze.

curator: A person responsible for the care, display, and study of particular collections, such as works of art.

cyanotype: A photographic print that shows white outlines on a blue (cyan) background, a product of the chemicals used on the print's photosensitive paper.

Daguerre, Louis-Jacques-Mandé: (French, 1787–1851) The inventor of the photographic process that bears his name (daguerreotype).

daguerreotype: A founding method of photography, announced in France in 1839, in which an image is fixed on a silver-covered copper plate made sensitive to light.

decomposer: In an ecosystem, an organism that breaks down waste materials from producers and consumers.

dendrochronology: The study of growth rings in trees, for measuring environmental variations and dating wood.

direct lost-wax method: A technique used to cast metal sculptures. In this process, a clay model is created first. Next, the model is covered with a layer of wax and then a much thicker layer of plaster. The entire mold is heated to melt the wax, and molten metal is then poured in to replace the "lost" wax. After cooling, the mold is removed to reveal a metal sculpture that preserves the details sculpted in the original clay model.

ductility: The ability of a metal to be easily manipulated into a form (by stretching, for example).

ecosystem: A community of organisms in a specific environment.

element: A substance that cannot be chemically broken down into simpler components.

enamel: A protective and/or decorative coating baked on to a ceramic, glass, or metal object.

energy: Usable power that allows work to occur through the application of force.

entomological: Related to entomology, the study of insects.

evaporation: The process by which a liquid is converted to a gas.

ewer: A decorative pitcher with a base, body, and spout.

exoskeleton: An external skeleton that supports and protects the body of an insect.

faience: A type of earthenware covered with thin, opaque layers of colored glazes.

flat slide: A small glass or plastic plate on which a specimen is placed for viewing under a microscope.

fleur-de-lis: In French, the "flower of the lily," a stylized iris design used in art and architecture as the symbol of French royalty.

food web: In an ecosystem, a series of interconnected food relationships among producers, consumers, and decomposers.

force: The power exerted on an object that influences its condition, motion, or position.

genre: A style of painting with a specific type of subject matter, including history, portraiture, landscape, still life, flower painting, and scenes of everyday life.

Giambologna: (Flemish, 1529–1608) Influential sculptor, active in Italy between 1550 and 1600, who was one of the leading proponents of Mannerism.

gilding: A surface, such as on furniture or other decorative arts objects, covered in gold leaf or gold paint.

glair: A substance made from egg white, used to prepare a surface for decoration and as a binder.

glass negative: A glass plate coated with photosensitive material and used for capturing an image in a camera.

glass paste: Glass that is ground up and combined with an adhesive in order to decorate objects such as jewelry, statues, and furnishings.

glaze: A liquid substance added to the surface of ceramic objects that creates a protective and/or decorative coating when fired.

grotto: A structure built to resemble a cave.

Guild of Saint Luke: A seveteenth-century Dutch trade organization for artists and artisans, responsible for training young artists and regulating the artistic crafts.

guilder: Currency of the Netherlands from the thirteenth century until 2002; derived from *gulden* (golden).

habitat: The environment in which a plant or animal usually lives.

head: The anterior part of the body of an insect, containing the eyes, antennae, mouth, and brain.

heat: A form of energy applied to a substance that changes its temperature, form, composition, color, condition, etc.

Hellenistic: Related to the art and architecture of Greek culture from 323 B.C. (the peak of influence after the conquests of Alexander the Great) to 146 B.C. (the defeat of Greece by Rome).

herbivore: An animal that feeds on plants.

Hertzsprung-Russel (HR) diagram: A two-dimensional graph created by Ejnar Hertzsprung and Henry Norris Russell about 1910 that plots the temperature and luminosity of stars, which allows for the analysis of stellar evolution.

Hoefnagel, Joris: (Flemish, 1542–1601) Self-taught artist who was a pivotal figure in the history of art from the Netherlands, both as the last important Flemish manuscript illuminator and one of the first artists to work in the new genre of still life. He served as court artist to Albert V, duke of Bavaria, and the Holy Roman Emperor Rudolf II, creating numerous natural history miniatures for books.

Hooke, Robert: (English, 1635–1703) Experimental scientist, active in the fields of astronomy, biology, chemistry, and mechanics, among others, who published his discovery of plant cells in *Micrographia* (1665).

horticulturist: A specialist in the art and science of growing plants.

Huguenot: A French Protestant of the sixteenth and seventeenth centuries, often persecuted because of opposition to the Catholic Church.

illumination: The art of decorating manuscripts with colored and gilded initials, borders, and miniatures.

insect: A small invertebrate with a body divided into three parts—the head, thorax, and abdomen—with three pairs of legs, generally one or more pairs of wings, and a pair of antennae.

invertebrate: An organism lacking a spine.

isotope: Forms of the same chemical element, but with different atomic mass and physical properties.

kaolinite: A mineral that is the chief component of kaolin, an ingredient used for making porcelain and other ceramics.

kiln: A heated enclosure for firing ceramic works.

lapis lazuli: A bright blue semiprecious stone that is ground to make a pigment of this color.

lens: A piece of glass or other transparent substance with curved sides used for magnification or focus.

lifespan: The length of time an organism is alive.

light: Luminous energy from a natural or artificial source that stimulates the process of sight.

light wave: The movement of light particles.

line: A mark with greater length than width, which describes a contour, establishes a boundary, creates a design, or otherwise defines space; one of the elements of art.

Linnaeus, Carl: (Swedish, 1707–1778) Naturalist and botanist who established the modern method of classifying plants and animals in such publications as *Species Plantarum* (1753).

liquid: A substance that is neither a solid nor a gas, in that its molecules flow freely (unlike a solid) but do not tend to dissipate (unlike a gas).

madder: A moderate to strong red pigment made from the roots of a European herb of the same name.

majolica: A type of earthenware decorated with opaque glazes made from tin oxide.

malleable: The ability of a metal to be easily hammered or pressed into a form.

mandible: Part of the mouth of an insect, used for holding or biting food.

Mannerism: Stylistic phase in European art between the High Renaissance and the Baroque (roughly 1510 to 1600), characterized by exaggerated compositions with elongated forms and contorted poses.

manuscript: A document handwritten on parchment, leather, or other material, necessary before the introduction of mechanical printing.

marble: A type of hard, often colored limestone used for architecture and sculpture; also, a sculpture made from this material.

marshland: An ecosystem of wet, low-lying land with grassy vegetation.

Medici: A powerful Florentine family, active in banking, commerce, politics, religion, and art patronage, from the fourteenth through eighteenth centuries.

metal: A chemical element, such as gold or iron, with properties of conductivity, malleability, fusibility, and ductility.

Metamorphoses: A narrative poem by Ovid that details the transformations (metamorphoses) of characters in Greek and Roman legends and myths, beginning with the creation of the world.

Michelangelo: (Italian, 1475–1564) High Renaissance/Mannerist sculptor, painter, draftsman, architect, and poet, responsible for some of the world's most recognized works of art, including the statue *David* (1504) in Florence and the painted ceiling of the Sistine Chapel (1512) in Rome.

Micrographia: A text written and published by Robert Hooke in 1665 on the topic of biological specimens, including cells, viewed through a microscope.

microscope: An optical instrument used for magnifying specimens, such as cells, that are too small to be seen with the human eye.

mineral: Inorganic matter, neither animal nor vegetable, that occurs naturally in earth and water.

miniature: A small, detailed painting in an illuminated manuscript (from the Latin *miniare*, "to paint red," referring to the red pigment [minium] used for paint and ink).

mixture: A chemical combination of two or more elements that do not bond together.

molecule: A group of atoms bonded together; the smallest unit of a compound that can take part in a chemical reaction.

Montmorency, Anne de: (French, 1493–1567) Soldier and duke who led Catholics to several victories over Protestants in the Wars of Religion.

mount: An ornamental piece, usually made of gilt bronze, attached to furniture and ceramics as both decoration and protection.

mull: To grind into a powder or beat until well mixed.

negative: The reversal of dark and light tones of a photographic image, captured on various media and used to create a positive print.

Newton, Isaac: (English, 1642–1727) Mathematician, physicist, and astronomer, who invented calculus, studied light and color, and described, through his theory of gravitation, how the universe is held together.

Newton's third law of gravity: Isaac Newton's principle that "for every action, there is an equal and opposite reaction."

ocular micrometer: A ruled scale in the eyepiece of a microscope, used to measure the size of magnified objects.

omnivore: An animal that feeds on flesh and plants.

Ovid: (Roman, 43 B.C.–A.D. 17/18) The author of *Metamorphoses* (A.D. 1–8) but also of intelligent and humorous love poems, one of which, however, resulted in his expulsion from Rome.

oxidation: The process by which one chemical compound loses electrons to another, as when metal rusts (the metal reacting with oxygen in moisture).

oxidation-reduction reaction: The process by which electrons are transferred between the atoms or molecules of different chemical compounds, with the compound losing electrons being oxidized, and the compound gaining electrons being reduced.

oxide: A compound of oxygen and another element.

oxygen: A chemical element, essential for life, that is a component of air, water, and earth.

patina: Green or brown coloration on the surface of metals, either applied or caused by oxidation.

pattern: A repeated motif, element, or design.

perpendicular: Being at a right angle to something.

photograph: A fixed image captured by using a camera.

photography: The method of capturing an image by chemical, mechanical, or electronical processes.

photosensitive: The quality of a material that makes it react to light.

physics: The study of matter and energy and their relationships.

pigment: A powdered substance mixed with a liquid to add color, as in paint or ink.

porcelain: A hard, white ceramic material made of kaolinite clay fired at high temperatures; often coated with glazes.

press-mold: To shape an object by pressing its source material into a mold; also, a mold used for this purpose.

primary consumer: In an ecosystem, an organism that feeds on plant material (such as a rabbit eating grass).

producer: In an ecosystem, an organism that produces materials consumed by other organisms.

provenance: The history of ownership of a work of art.

pull: To exert force on an object in order to move it toward the source of the force.

push: To exert force on an object in order to move it away from the source of the force.

radiation: The process by which heat is transmitted through a substance in the form of waves.

raw umber: A pigment made of, and the color of, natural earth.

reflective: Capable of casting light back to a light source.

refraction: The deflection of light, radio, and other types of waves as they pass between mediums of different densities.

Renaissance Revival: A nineteenth-century style in art and architecture that was influenced by earlier styles that drew inspiration from ancient Greece and Rome.

Rococo: An elaborate decorative style of painting, architecture, and interior decoration, characterized by asymmetry and natural motifs, originating in France and becoming popular throughout Europe from the 1720s through the 1760s.

salted-paper print: The first type of paper print used in photography, whereby a positive image was created by placing photosensitized printing paper under a negative and exposing both to light, then coating the resulting print with a salt solution in order to fix the image.

Sambin, Hugues: (French, about 1520–1601) Wood-carver, designer, architect, and engineer who, in 1551, became master of the guild of *menuisiers* (designers) and supervised many design projects for Charles IX, king of France.

Savery, Roelandt: (Flemish, 1576–1639) Draftsman, painter, and printmaker whose work for the Holy Roman Emperor Rudolf II included studies of animals in the emperor's menagerie. Savery's creations played important roles in the development of several genres: floral still lifes, paintings of cows and other animals, cityscapes, and landscapes.

scientific illustration: A representation of something from nature that captures its details with accuracy.

secondary consumer: In an ecosystem, an organism that feeds on primary consumers (such as a fox eating a rabbit).

semi-permeable membrane: The part of a cell surrounding the nucleus that allows some substances, but not others, to pass through.

shape: An enclosed space formed by a closed line; one of the elements of art.

silica: A nonreactive chemical compound, naturally occurring in such materials as sand, that is ground and mixed with pigments as a component of glazes.

Snyders, Frans: (Flemish, 1579–1657) The first specialist in animal still-life painting, a new Flemish genre in the 1600s. These works allowed Snyders to display his skill at organizing a rich variety of textures, colors, and shapes. In addition to his own energetic hunting scenes and complex still lifes, Snyders was often employed by his close friend Peter Paul Rubens on the still life and animal sections of Rubens's paintings.

solid: A substance that is neither a liquid nor a gas, but rather firm in shape (unlike a liquid) and stable in condition (unlike a gas).

solution: A chemical mixture where one or more elements is dissolved in another element.

species: A category of organisms with similar characteristics, capable of interbreeding in nature and thus passing along genes.

specimen: A sample of something used for scientific study, or an example of something used as a representative of a particular species.

tensile strength: The maximum tension a material can bear without tearing or breaking.

tertiary consumer: In an ecosystem, an organism that feeds on secondary consumers (such as an eagle eating a fox).

thorax: The middle part of the body of an insect, between the head and the abdomen, bearing the legs and wings.

tin: A chemical element, with anti-corrosive properties, used in making alloys such as bronze.

Ursa Major: A constellation perceived as resembling a "great bear" (in Greek myth, the nymph Calypso, transfigured by Zeus).

value: Gradations of light and dark; one of the elements of art.

varnish: A resinous liquid applied to a surface to provide a transparent protective coating.

vermilion: A bright red pigment made from the mineral cinnabar or from the reaction of mercury and molten sulfur.

vertebrate: An organism possessing a spine.

vitrification: The process of converting a material into glass or a glasslike substance by applying heat.

Wars of Religion: A series of wars fought in France (1562–98) between Protestants (Huguenots) and Catholics over religious doctrine.

wash: A thin coat of watercolor or ink applied with a brush to a paper surface.

waxed-paper process: A photographic process in which a paper negative is coated with a layer of wax prior to exposure and development, allowing for delayed use of the negative and greater clarity of the image.

Wollaston, William Hyde: (British, 1766–1828) Chemist and physicist who invented the camera lucida in 1807.